DATA FROM NATURE

NATURE, DIGITISATION AND DESIGN

CONTENTS

ACKNOWL-EDGEMENTS

What began as a private passion has grown to involve many others: I would like to thank the following, and apologise for any inadvertent omissions.

My first assistant, Sophie Shardelow, played a vital role by insisting that I attend the Liberty 'Open Call'; her father Alan did sterling work in preparing my 'laboratory'/house for the arrival of the television crew and later assisted with various domestic projects and exhibitions.

Martin Price's extensive business experience was invaluable when the breakthrough with Liberty came and he now runs Weston Earth Images with his wife Helen.

Various others have contributed as full- or part-time assistants, including Ceri Redman, Amanda Williams, Rosie Blatchford and Francesca De-Lilla (whose mastery of pattern-making using CreateForAll is outstanding). Manisha Harkins' background as a journalist-turned-textile-designer enables her to turn her hand skilfully to all aspects of the current range of work and a new journalism graduate, Caterina Dassie', is taking charge of Molly's 'Video Correspondents'.

I thought being on a prime-time BBC television series might lead to a flood of interest (!), but it produced only two enquiries of consequence, from Russ Billington and Judith Mackay-Solanki. Russ generously provided vector tracings of Arts and Crafts and other patterns, while Judith, a highly gifted jewellery designer, experimented with silk-in-resin and other costume jewellery: commercially her efforts sadly led nowhere, but creatively her enthusiasm was infectious.

I am enormously grateful to Patel Taylor Architects for their support and live in hope that others may follow their example.

My way into digital fabric printing came courtesy of Peter Cattermole, who later introduced me to Doug and Elaine Davies of the Silk Bureau, and recommended I contact Renzo Veronelli of Maver in Como. Renzo's flexibility in supporting a small company making frequent requests to try out ideas beyond the core business of scarves was much appreciated.

Creating apps took me into yet more uncharted territory. I am grateful for the coding skills of Joe Offside, and for the support of Angela Lepore, now Wil Howlett, and Steve Rees, head teachers respectively of Albany and Evenlode primary schools, where we continue to test possibilities.

At Cardiff University, Professor Dianne Edwards FRS and Dr Alan Channing of the School of Earth and Ocean Sciences provided support and encouragement, and the latter has kindly allowed me to incorporate parts of a text he wrote for an earlier book. Following early retirement I was delighted to be appointed Visiting Professor of Surface Design at Cardiff Metropolitan University: particular thanks to Keireine Canavan and her textile design students.

Over the years I have formed valuable practical relationships, including Dave Stafford of Beacon Printers; Threadster fabric printers; Jammy Custard Animation; Spencer Smith of Nikon Metrology; and, above all, Deborah and Simon Campbell whose shop, Debris, was a valuable source of specimens and whose enthusiasm helped keep me going.

Finally, I would like to thank unnamed friends and my fellow directors of Molly's World CIC, John Parker and Rebecca Pike, for personal and practical support, and Andrew Mead for editorial and moral support over many years.

Picture and other credits

The photographs of Patel Taylor's work are by Peter Cook and that of Amy Churchlow on p.140 by Kate Williams. The dresses on pp.114-5 are by Eden Diodati and that on pp.116-7 by Olga Alpatova; the lamps on p.122 are by Mark Ebbutt and the silk-in-resin bangle by Judith Mackay-Solanki. The images of ceramic glazes on pp.86-91 are from handmade dishes by Simon Rich and specially prepared tiles by Geoffrey Swindell; those of glass on pp.92-7 were made from variously distressed specimens created by Rodney Bender of Innovative Glass Products. The painting of The Last Judgement on p.179 is at Uppsala University; the cabinet of curiosities on p.181 is in the Metropolitan Museum New York; and the picture of Perseus Liberating Andromeda on p.182 in the St Louis Art Museum.

"There are no natural objects out of which more can be learned than out of stones. ... For a stone, when it is examined, will be found a mountain in miniature. The fineness of Nature's work is so great, that into a single block, a foot or two in diameter, she can compress as many changes of form and structure, on a small scale, as she needs for her mountains on a large one; and, taking moss for forests, and grains of crystal for crags, the surface of a stone, in the plurality of instances, is more interesting than the surface of an ordinary hill; more fantastic in form, and incomparably richer in colours."

John Ruskin, Modern Painters, Vol. IV

Above: image captured from a few millimetres of an agate.

FOREWORD

AN ODYSSEY IN MINUSCULE WORLDS

Do the objects of our passions attract us as much as we are captured by them? Professor and Architect Richard Weston's adventures in the tiny world of minerals brilliantly give this impression.

His odyssey started with the magnetic attraction for a *sleeping* object. It is easy to imagine him striding around the streets of Cardiff, rush into one of the city's covered arcades, then stop, his gaze captured by the beautifully shining reliefs of an ammonite patiently lying in a shop window. The fossil ends in the walker's bag as quickly as it was noticed. From this moment, he remembers all the missing colours and sparkles that can be seen during a more careful look with the scanner: fifteen years later, the *fateful ammonite* has led the Professor to a properly equipped visual research which has yielded thousands of amazing pictures of materials, minerals and rocks.

During many years the Architect patiently composed a collection of beautiful microscopic curiosities from opals, glasses, limestone and paint. Among all the materials contemplated through the microscope, the agates are his favourite. Maybe because all their morphological complexities have not completely succumbed to the scientists yet, but surely because their bands and coloured bubbles evoke fabulous infinitesimal universes. The Architect is captured by the clues hidden in the rocks, related to their past sudden appearance. His eyes are delightedly plunged into mazes of millimetric shapes and shades in the quest for elegance on the brink of what can be seized. They reveal tsunamis, placid seas, paradisiac islands or baroque foams. Weston's pictures display tiny worlds animated by a myriad of events undetectable to the naked eye, filled with sleepy souvenirs that can only wait to be kindled.

Richard Weston never tires of revealing nature's forms and subtleties, a guilty pleasure that fits in a tradition which it is not exaggerated to trace back to the origins of architecture. From ancient columns to today's Biomimetics nature acts at times as source of inspiration, at times as model, starting from its mythical origins: in *Eupalinos or the Architect* by Paul Valéry (1921), Socrates almost becomes an artist or architect at the sight of a strange object shaped by the sea. The historian behind the architect cannot altogether help himself from finding similarities between his discoveries and the different fields of Art. Some specimens of pyrite reminds him of a sketch from deconstructivist Peter Eisenman; another mineral resembles a Malevitch painting instead. Professor Weston has also isolated several hundreds of graphic images that he ascribes to a human hand, that of Carlo Alcite. The work of this obscure Italian master of the XVIIth century, whose life has long gone unnoticed, might soon be revealed by Weston in a biography. Do these images represent the reconstruction drawings of the Roman aqueduct in Genoa dating four centuries back or photographs of calcite? The mystery remains.

Ten years ago, Weston's collection of precious pictures took a turn that he himself could have never imagined. His microcosms of colours and lights got away from the Welsh atelier to propagate on scarfs, foulards and umbrellas, as well as smartphone cases that could be found at the *Bon Marché* stores in Paris, while being published in the world's most famous fashion magazines. The origins of this unusual trajectory lie in a TV show. The Architect, Professor at Cardiff University and author of more than 15 scholarly books, besides being a great expert on Modernism and the works of Alvar Aalto and Jørn Utzon, was among the finalists of a talent show putting together 600 participants. Quite naturally, his tiny worlds have appeared in the guise of multiple avatars: pictures of variscite, moonstone or paesina stone that colonise the flagstones and tiles paving his garden and house floor; hangings decorating the curtains and carpets; and wall panels or patterned glass used for broader scale parts of the structure. Recently, this non-stop change of dimensions – from 3D to 2D and 3D again – have finally brought to an ultimate transmutation: by moving to virtual reality, amateurs of the *tiny worlds* can now walk across a one-centimetre chalcedony island.

Born in 1953, Professor Weston is hardly a *digital native*. Back in the years 2000, before his captive encounter, he still used to teach his students with the aid of diapositive slides. It is therefore paradoxical that a timeless fossil has led him to convert to modern technologies, without leaving behind his academic research: the collection of visual gems allows him to pursue at the same time his investigations on the beauties of nature and its reception. Weston is astonished by how deeply sensitive to the images of natural patterns are those who contemplate them, whatever their age and origins. Not so long ago, Professor Weston encouraged his students to "learn" about nature in project workshops; now, his microscopic landscapes are in the hands of children less than 10 years old.

Today, Weston spends most of his time on *Molly's World*: a world in which his late cat is a digital character that evolves with Patch the dog. The two friends wander around at the youngsters' expert 'fingers' discretion, busy filling up the world they travel in with pixels of colour, all coming from the vast collection of textures drawn up from the Architect's visual research.

In this pocket-size galaxy, pictures of calcite or fluorite become pieces of art, design patterns, clothes for animals or just colouring pages, depending on the mood. The designer's experiments also continue before our eyes. Recently, while engraving a William Morris's *Strawberry Thief* pattern on a slate, the laser beam accidentally brought to the boil the water contained in the rock, leaving pyrite to escape and subsequently recrystallize on the surface. The animal and flower drawing became decorated with a golden relief.

Richard Weston has the skills to slide between materials and scales. By transforming minerals and crystals into marvellous landscapes, he has opened a creative gap between gold panning and alchemy, where intuition, science and luck become entangled with a touch of magic that could enable a garden thrush to wear golden wings.

Béatrice Durand
Architect, Researcher and Journalist

INTRO-DUCTION

In the autumn of 2003 I was walking through one of Cardiff's old arcades and in the window of a shop called 'Crystals' noticed a large and very beautiful ammonite, glowing with golds, deep reds and an array of subtle greens and umbers. At the time, I had been enjoying digitally scanning assorted natural things – leaves, feathers, tree bark – brought back from walks outside my village. I decided to buy the fossil to see what delights it might yield, and was bitterly disappointed when all my scanner could capture was a muddy brown shadow of what the unaided eye could see.

Reflecting on the fact that the scanner cost less than the fossil I decided to buy one of the best Epson had to offer. The results were a revelation. Filled with what I later learned were differently-coloured mudstones, the once empty chambers were alive with flecks of golden calcite. I showed a print of a detail to a colleague, introducing it to him as a terrazzo floor compartmented by serpentine strips of brass. 'It's obviously Italian,' came the response, 'but I don't recognise it.'

The work with 'data from nature' presented in this book flows directly from this encounter with what I came to call 'The Fateful Ammonite'. For the next ten years I continued to teach, and write about, architecture, but increasing amounts of my spare time – and money – were devoted to 'the scans'. I was captivated by the sheer beauty of the images that many minerals yielded and, initially, had no thought of possible applications for the steadily growing library of digital files. A couple of years into the endeavour a friend suggested that, printed onto fabric using then new digital techniques, some might well find a place in interior decoration or fashion – as they eventually did, in Liberty of London's famous scarf hall.

It then struck me that small areas of the high-resolution images also offered a means to 'digitally paint' drawings and patterns. It was some time before I realised that this was, in effect, a digital reinvention of the process of making pictorial inlays using decorative minerals and stones – *pietre dure* as they are known in Italian – that was raised to astonishing heights of intricacy and verisimilitude under the patronage of the Medici in seventeenth-century Florence.

After turning sixty in 2013 I decided to take early retirement from the university, where my curious private passion had no place in the Research Excellence Framework by which academics' work is now evaluated. Convinced that many of the minerals held delights beyond eye or scanner, I used part of my 'pension pot' to buy a powerful Nikon microscope: the revelations that flowed from its digital camera exceeded all expectations.

As will be clear from this brief summary, the fifteen-year-long engagement with data from nature presented here has been anything but pre-planned, and in organising this book I have resisted the temptation to impose a seemingly more systematic structure. Many of the pleasures it has afforded have come from chance discoveries and pursuing tempting byways, and this is reflected on occasions within the chapters.

The book begins with a brief discussion of the science of minerals, into which I have delved along the way with the help of a 'geological friend', Dr Alan Channing. This is followed by an extensive presentation of the full range of the 'data from nature', including examples of 'honorary nature' like squeezed paints, ceramic glazes and an unexpected by-product of car manufacturing known as Fordite. Chapters 3, 4 and 5 present applications ranging from fashion and architecture to digital apps and the 'laboratory' of my own house and garden. We then step briefly aside to review some of the historical pleasures of minerals, including people's long-standing fascination with discovering in them assorted images ranging from credible landscapes to implausible portraits of saints and historical figures. The final chapter presents a sequence of such 'pareidolic' images: captured with the microscope from the surfaces of seemingly featureless calcite crystals they are so vivid in their evocation of varied topographies that I assign them to a lost artist, Carlo Alcite.

Throughout these explorations I have felt what can best be described as a growing kinship with minerals. When I showed several early prints, including this image (next spread) from a piece of jasper to a neighbour, he declared spontaneously, 'Everyone's going to love these Richard.' I asked him why and he said, 'Look at this one, you've got the cosmos at the top and the beginnings of cellular life at the bottom. *It's made like us*.' In the Conclusion I reflect briefly on this remark, 'made like us', relating it to the ideas of John Ruskin and to a growing body of ecologically-motivated thinking that is developing around the idea of 'vibrant matter'.

1

MINERALS
IN
NATURE

Think of the word 'mineral', still more 'crystal', and what almost certainly comes to mind are the hexagonal shafts and pyramid-like points of quartz, or the intersecting cubes of pyrite that appear, to an architect, almost like studies for Modernist buildings. The possession of a geometric, crystal structure is one of the defining features of minerals but for the most part in nature it is manifested at a microscopic scale. Most grains of sand are quartz, and while cubic crystals may be the molecular form of pyrite, it is most abundant as what mineralogists called framboids. Derived from the French word for raspberry, these clusters of tiny spheres are generally very small (typically 0.01mm); ancient (they have been found in rocks in South Africa aged at 2.9 billion years); and staggeringly numerous. Informed guestimates put their number at 10^{30}, which is 10 billion times the number of sand grains on Earth or about a million times the estimated number of stars in the universe.

I have hundreds of calcite rhomboids – archetypal crystals with exactly the same angle (74°55') governing their shapes: they are the source of the 'work' of Carl Alcite discussed in chapter 7. But I also have large clusters of planar crystals and small hemispheres that are also calcite, just three of over eight hundred known forms of the same mineral. What determines their large-scale formations are a range of other factors affecting how they grew. A common form of crystal growth, known generically as botryoidal (named by Pliny the Elder after the Greek for grapes), consists of a clump of globular formations formed by linear crystals growing around a crystal seed or speck of sand, dust or other material. A perfect example is malachite, the green mineral found in association with copper: to the amateur, the intersecting green spheres hardly appear crystalline at all, but a polished slice instantly reveals the beauty of the growth of the crystals within.

Scientifically speaking, minerals are precise assemblages of chemical elements that bond with a fixed crystal structure. Some, like the ubiquitous quartz, involve only two elements, silicon and oxygen, bonded in a continuous matrix of neat tetrahedra, while others are truly exotic and found in vanishingly small quantities: cobaltarthurite, for example, comes with the imposing composition of $(Co,Mg)Fe^{3+}_2(AsO_4)_2(OH)_2 \cdot 4H_2O$ but it is thought that all there is of it on Earth would fill no more than a thimble. New minerals are regularly being discovered – up to 40 or 50 a year – while a few others are 'demoted' from official recognition, so the official numbers are surprisingly imprecise: various apparently authoritative sources put the figures at 4-5,500.

The earliest known efforts to describe minerals quasi-scientifically were made in ancient China. The challenge also exercised the mind of one of Aristotle's most distinguished followers, Theophrastus, whose ideas influenced their treatment in Pliny the Elder's celebrated *Natural History*. Pliny was the first to appreciate that amber formed from a viscous liquid exuded by certain trees, and that diamonds share an underlying octahedral structure, but his knowledge was inevitably unsystematic and fragmentary. He also seems to have been the first to relate appearances to inner structure, wondering why, in the case of quartz, nature should choose to 'build with hexagonal bricks'. Quartz crystals were then widely supposed to be a form of deeply frozen pure water: the recognised 'elements' from which everything was made were, of course, only four – earth, air, fire and water. But despite this insight Pliny nonetheless assumed that the complex crystalline gemstones brought from India were the work of craftsmen, not polished versions of natural forms.

The beginnings of mineralogy as a science was marked by the publication in 1546 of *De Re Metallica* by the German physician Georgius

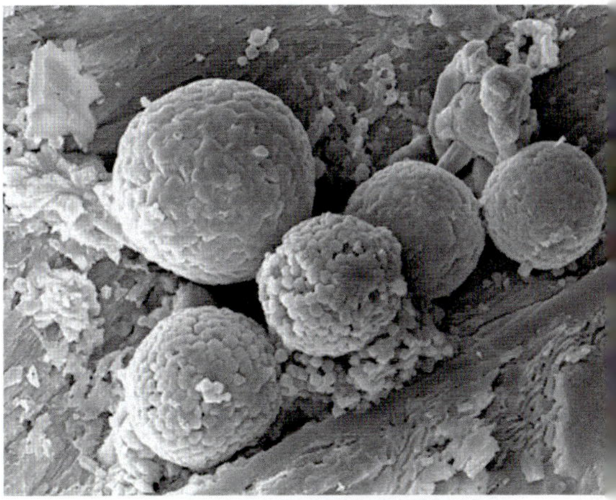

Agricola. In 1669 Nicolas Steno – who was also the visionary founder of the study of the layering of rocks, stratigraphy, in geology – examined minerals from many localities and noted that the angles between their faces were constant, regardless of their place of origin: this key insight became known as Steno's Law.

The science of identifying the mineral constituents of rocks is known as petrography and began to be placed on a secure footing in 1828 when the Scottish physicist William Nicol invented a technique for producing polarised light. This involved cutting an optically clear crystal of Iceland spar (the 'calcite rhomboids' referred to above) into what became known as a Nicol prism. Two such prisms, when rotated, produce polarised light, and this can be used to determine the crystallographic character of tiny mineral grains. During the 1840s the technique was greatly advanced by the making of what are known as 'thin sections' of rock. This involves gluing a slice of rock to a microscope slide and then grinding it down so that it is sufficiently thin to allow light to be transmitted through mineral grains that otherwise appeared opaque. 'Thin section petrography' is now learnt by all student geologists: with the thickness standardised at 30 microns (thousandths of a millimetre) and full (90°) cross-polarisation the minerals reveal signature colours and features. For the hunter after data from nature, strikingly different colour palettes may be revealed from the same material by varying the degree of polarisation.

Using traditional chemical techniques, determining the composition of minerals was a laborious process and understanding the

crystal structure all but impossible. Nineteenth-century mineralogists spent much time measuring the angles between crystal faces with great precision, but certainty did not come until the advent of X-ray crystallography in the early twentieth-century: the same technique that would later reveal the most important structure in nature, the double-helix of DNA, is also used to discover mineral structures that are similarly determined by the geometry of atomic bonding. The basic possibilities are surprisingly few, cubes, hexagons and sheets being the commonest. Minerals may also form fibrous structures, as in asbestos, or long needles, often thinner than human hair: rutile is among the most common and frequently encountered as what mineralogists call 'inclusions' in quartz and other crystals.

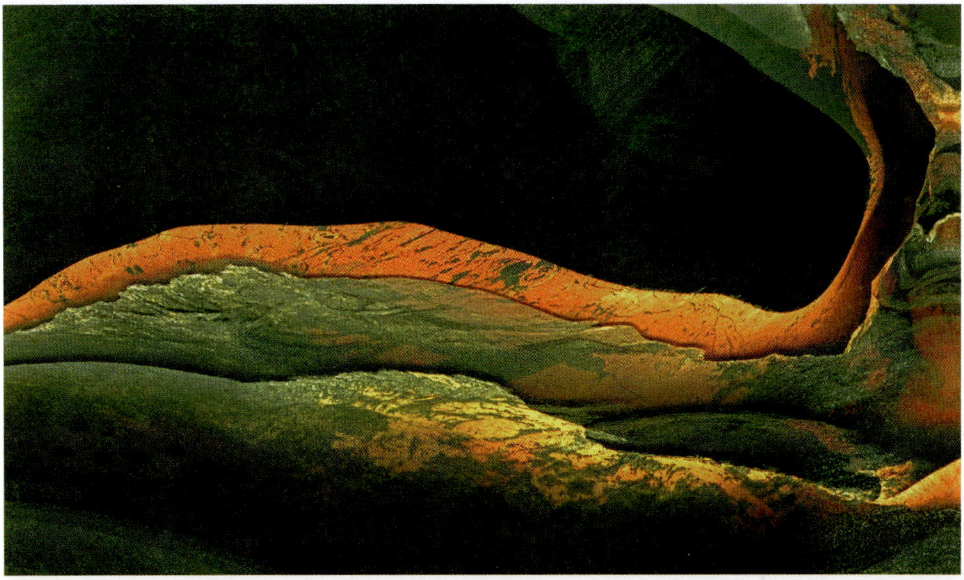

The underlying atomic structures create both the crystal faces we see and planes of structural weakness within the crystal lattice known as cleavage planes: these can offer rich pickings for the collector of visual data. Mica (right), for example, forms tabular, plate-like sheets which separate easily into layers far thinner than any paper: it behaves in this way because the chemical bonds between elements in the planes between the sheet surfaces are far weaker than those within the sheets themselves. Structural weaknesses – lines of cleavage – also occur within the plane of the sheets. They are distributed at 60° to one another and may be made visible by the movement of elements and minerals along them, especially when the mica is relieved of pressure and begins to delaminate – the source of the images on pp.66-69.

The determination of chemical composition was eventually made much simpler by the development of a relative of the electron microscope known as the microprobe. This came into widespread use in the 1950s and works by using electrons to non-invasively 'probe' cubic volumes as small as a few microns across. To the layman this sounds like an almost impossibly small amount to work with, until one remembers that it is sufficiently large to contain thousands of billions of atoms.

Large-scale mineral crystals are relatively rare in nature, but minerals themselves are ubiquitous, being the 'stuff' of which all rocks are made. Despite their apparent stability and permanence, rocks and minerals are in a process of perpetual flux, continually being created, destroyed, transported and

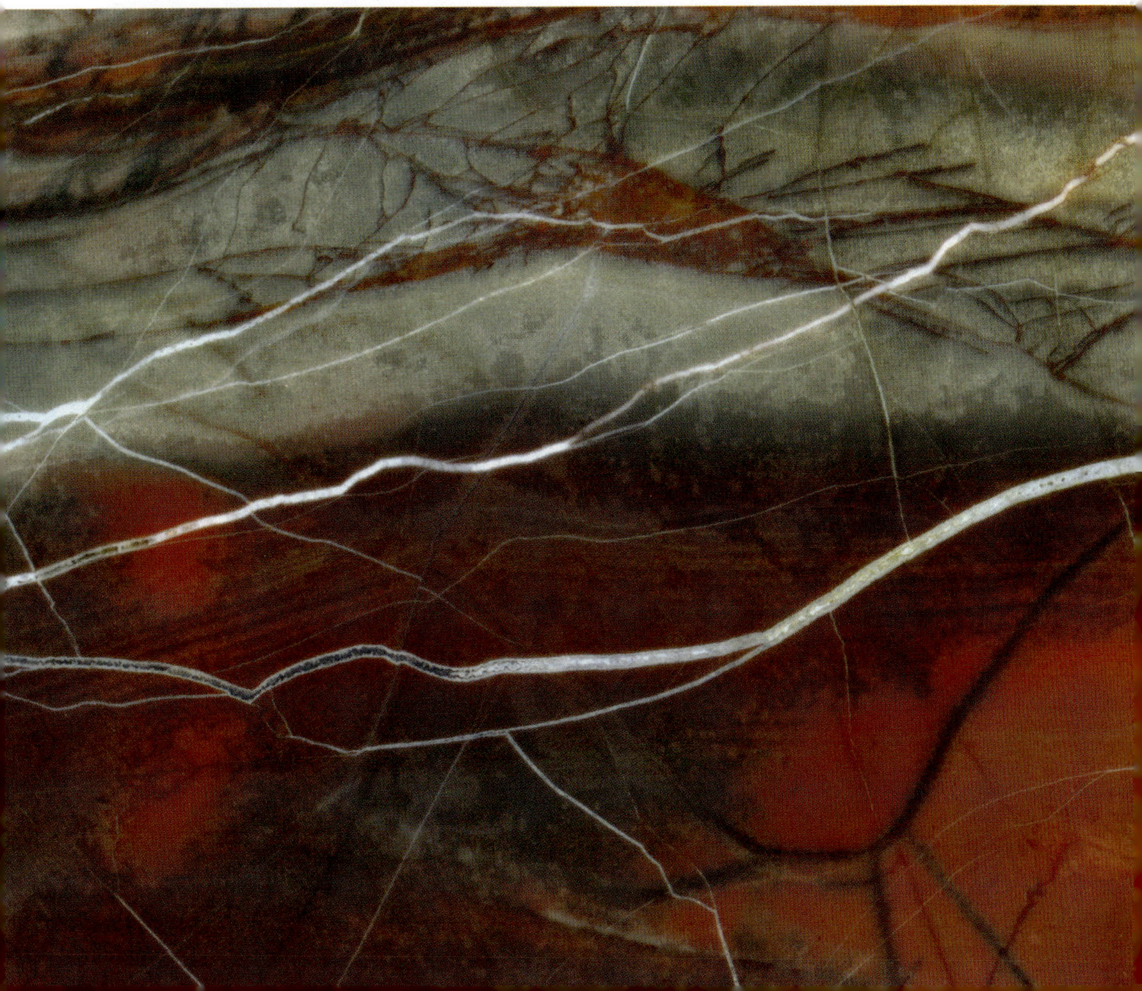

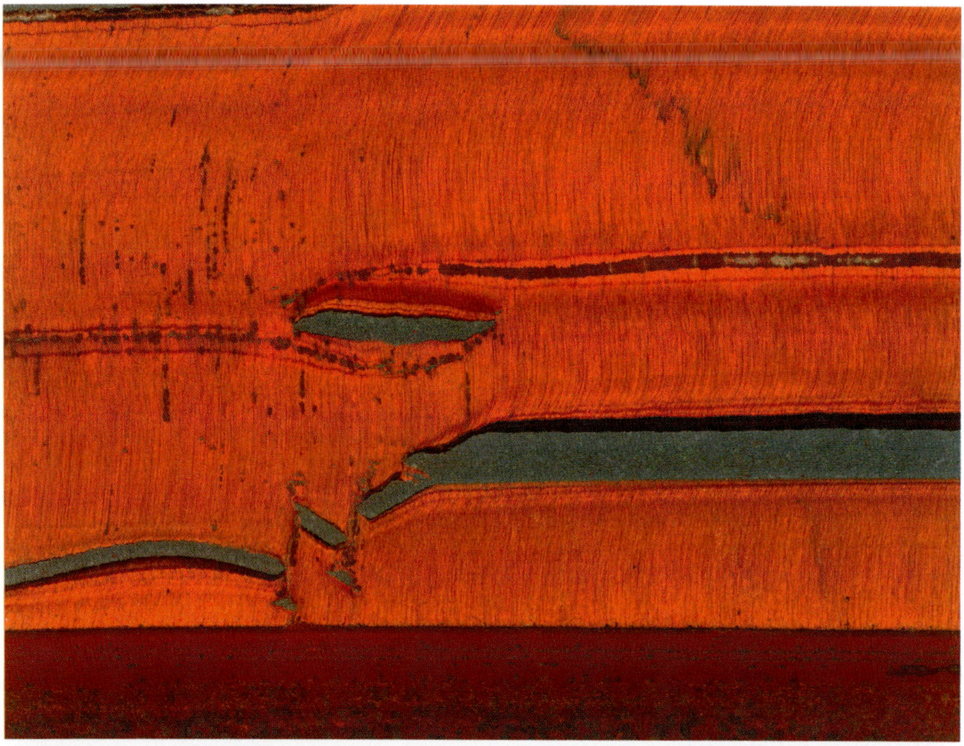

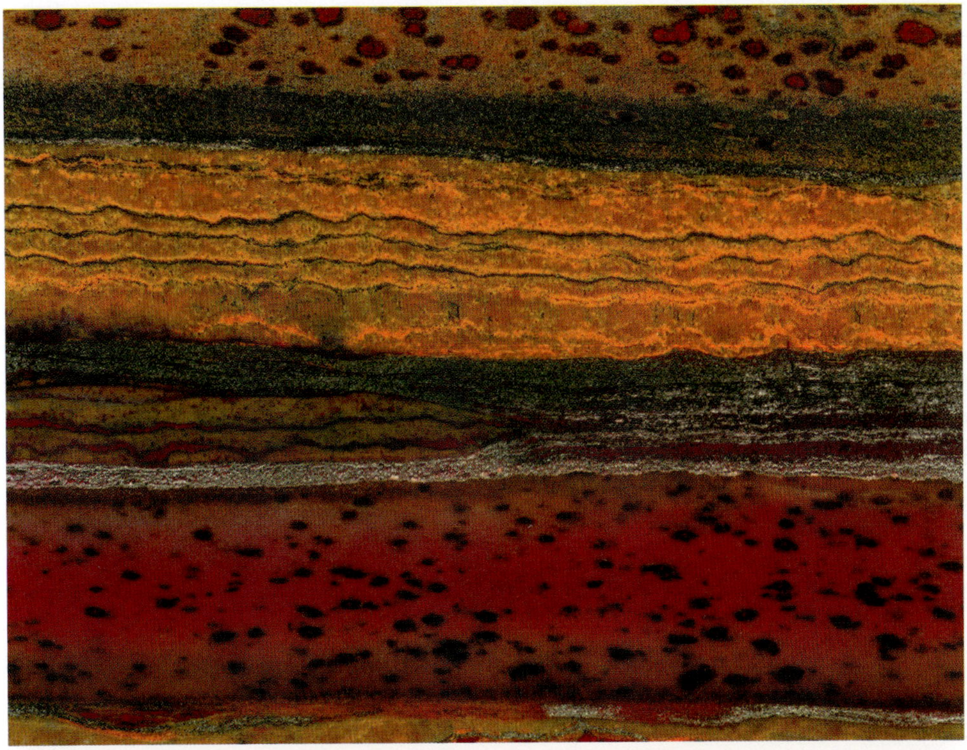

re-formed: geologists refer to this process as 'the rock cycle' and it is worth briefly exploring its essential dynamics.

Earth's basic structure is surprisingly simple – a solid nickel and iron inner-core; a molten outer-core made of the same elements; a solid but plastic mantle rich in magnesium and silicon; and finally a relatively thin but rigid crust divided into about a dozen major 'tectonic plates'. Through processes known as 'plate tectonics', rafts of cool, rigid crust are driven in a conveyor-belt-like fashion across the globe. Heat from the core and mantle is brought by convection to the Earth's surface, creating bodies of molten rock known as magma, in which many minerals form. Hot and buoyant, magma rises to the surface in linear chains of volcanic activity that traverse the centres of the major oceans. At these oceanic-ridges, magma erupts or intrudes into the crust and then cools to form igneous rocks.

Later generations of magma are forced up at the ridge, pushing the older rocks aside. Gradually this forces old, cold and therefore dense crust onto the oceanic-plate and ocean basin margins. Here, the dense crust descends back towards the mantle as it is forced below thicker and more buoyant continental crust in a process called subduction. Over millions of years, subduction can slowly devour entire oceanic plates, causing collisions between adjacent continental plates. When this happens, subduction cannot occur as both plates have similar buoyancy, and instead massive pres-

sures are exerted, forcing up mountain ranges like the Himalaya.

Beyond the crust, Earth's oceans and atmosphere act to destroy rocks, the next step in the rock cycle. Igneous rocks and minerals, created at high temperatures and pressures deep in the crust, are often inherently unstable in the cold, wet conditions of Earth's surface. Here they are chemically weathered, breaking them down to their constituent minerals or becoming altered, by hydration, to more stable minerals such as clays. Physical processes are also at work: ocean waves pound coastal cliffs, glaciers reduce obdurate granite to rock flour, and over aeons of time rivers carve vast canyons.

The water that provides the common agent of erosion is also a means of transport for rock particles. Glacial debris from the Himalaya, for example, is transported by vast river systems across the continents, eventually to be deposited in deltas and ocean basins. Particles of sand are driven thousands of miles across deserts by the wind. As the energy available for transport wanes, particles are deposited, forming unconsolidated sand, mud and clay sediments. Over time these build up layer by layer, bed by bed, and form thick sequences of sediments. These become progressively buried, and cementation and compaction transform them into solid sedimentary rocks like sandstones and mudstones.

Sedimentary rocks themselves are then eroded, potentially time after time, forming and reforming sediments and sedimentary strata. Sediment deposited in the ocean basins

hitches a ride atop the oceanic crust eventually to be transported to subduction zones and into continental collision zones. As oceanic crust and sediments are driven down subduction zones the fluids within them act to lower the melting point of the overlying continental rock. This generates partial melting and magma formation, creating volcanic activity. In collision zones, the sediments are compressed and heated in the process of metamorphism. Mud rocks and sandstones are compressed to form slates and schists. As the rocks are buried beneath up-thrust mountain ranges temperatures and pressures increase and they become plastic and deform, partially melting to create folded and banded gneisses and granites. The massive mountain ranges once again become zones of intense erosion and the rock cycle commences again.

Volcanism, metamorphism and deep burial in the crust create hot fluids that can dissolve rocks and minerals. Hot, buoyant, element-laden fluids then ascend towards the surface, progressively cooling along the way. The dissolved elements become saturated and are forced to precipitate by cooling and boiling (generated by reduced pressure), creating mineral veins and pods within older rocks – the sources of most of the 'data' in this book. Layer upon layer of minerals may be precipitated within a single vein as the chemistry of mineralising fluids changes over time, creating banded deposits, with mixed assemblages of minerals

like gold, silver, pyrite and lead. In other instances, where precipitation occurs slowly within open fractures, large, perfectly-formed crystals of minerals such as quartz can develop. The most spectacular recent discovery, reported world-wide, was of selenite crystals (a type of gypsum) up to 12 metres long and 4 metres in diameter discovered in the year 2000 in a large limestone cave in Mexico.

Played out at the scale of molecules or mountains – and beyond, at the planetary-scale itself – the molecular and geological processes we have briefly explored shape and re-shape the visible world we inhabit and the stuff of which

it is made. Many of the images in this book are graphical representations of points within this grand planetary-scale cycle. In a piece of obsidian (volcanic glass, p.32) we may sense a vivid expression of the volcanic flows from which it formed. A slab of sandstone (p.31) from Arizona a few centimetres across echoes the sedimentary layers of ranges of hills. Italian paesina stones (next spread) – greatly prized, as we will see later, by princely collectors – are criss-crossed by traceries of fine cracks caused by the collision between the African and European tectonic plates that created the Dolomites. A millimetre or two of other minerals may offer up, via a microscope, powerful images – formed millions of years ago and frozen in time – of geological processes, simultaneously creative and destructive and so hard to grasp because so infinitely slower than our time on Earth.

I will return to these thoughts in the Conclusion, but to end this all too brief survey of the science of minerals I would like to relate the first known attempt to explain in detail a crystal structure. Published in 1611, it took the form of an essay entitled *The Six-Cornered Snowflake* and was written by the great mathematician and astronomer Johannes Kepler as a Christmas present for his patron, Emperor Rudolph II. In it Kepler outlined possible explanations for the hexagonal structure of snow crystals – and then explained why each was wrong: 'seldom,' as a modern commentator has observed, 'has a mind of the first rank so charmingly described its *failure* to solve a problem.'

As an opponent of the atomic theory of matter, Kepler could not have reached a satisfactory explanation, and while the basic structure of snow crystals is now readily understood from the way water molecules crystallise and grow as ice, the symmetry of their 'arms' remains baffling. Snow crystals are frequently cited as examples of the endless variability of nature – no two ever being the same – and this variation is found as clearly, if more subtly, within individual crystals. Look closely at the examples shown here and you will see that the arms differ minutely in detail and are not neat examples of rotational symmetry. Snow forms while falling through the atmosphere, experiencing minutely changing environmental conditions such as temperature, humidity, air movement, contact with particulate matter: the resulting crystals are a product of internal (atomic) structure and the variability of the environment. The process may take only minutes in the case of snow, but the interaction of inner structure and environment holds true for the long durations of geological time.

Fascinating as the range of minerals may be, surprisingly few are the sources of the 'data from nature' that fascinate me. Among the several thousand known minerals, I have worked with only a few dozen, and more than 95% of the images come from varieties of quartz. These include agates and jaspers, both of which have many sub-species of their own. The differences between them often arise from impurities – which may be present at the outset or introduced during the process of formation – and variations in the process and environment of formation, which in turn depend on the geological context. The results are always unique, and while mineralogists can give a detailed account of the fundamental structure and composition of minerals, they are, for example, at a loss to explain adequately the processes involved in the formation of even the simplest-looking banded agates.

Crystals like quartz 'tips' might appear to be exemplars of the kind of 'Classical' beauty once thought to be grounded in geometry and proportion, and to permeate the universe in perpetual cosmic harmony. The dynamic view of the Earth, and of the Universe, constructed by modern science tells a very different and far more interesting story. As a source of digital 'data from nature' crystalline perfection would be of little interest: the minerals I work with are preeminent examples of that *other* beauty John Ruskin called 'vital', a beauty based not on imperfect shadows of static, ideal forms but at one with the modern understanding of the world as the product of flows and forces.

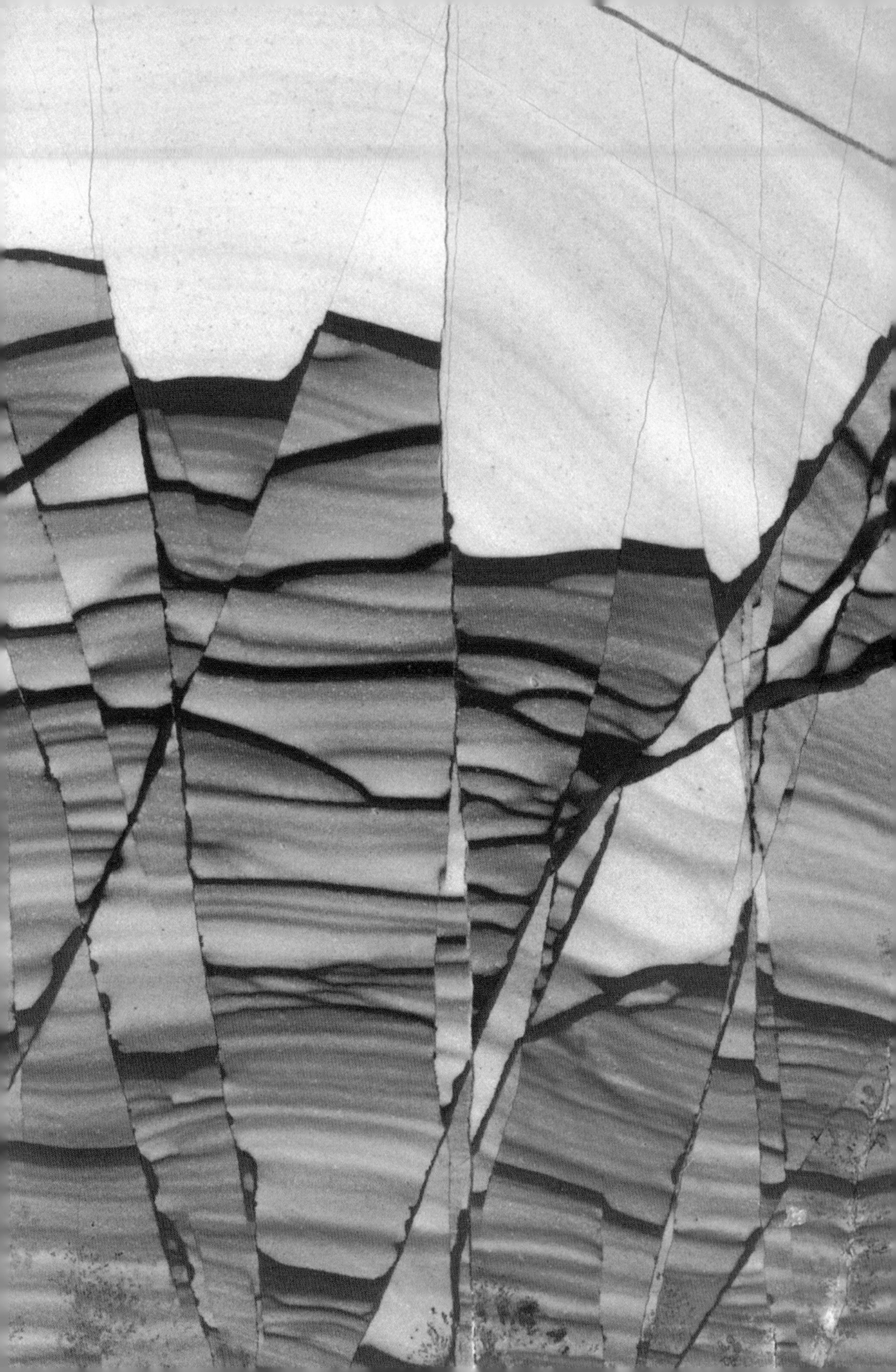

2

DATA FROM NATURE

Following the first, dismal attempt to capture data from 'The Fateful Ammonite' I bought an A3 Epson scanner with an optical resolution of 3,200 dots per inch and an adjustable focal plane that enabled me to resolve images up to a centimetre inside transparent materials, principally quartz. For the most part, however, I worked with flat polished slices of opaque or slightly translucent materials such as agate, the source of over a third of the files I saved. A difficulty with these slices is that they retain traces of the ultra-fine polishing powders used to bring them to a reflective sheen. Depending on the quality of the workmanship and the graininess of the material this appeared as anything from a scatter of isolated white dots or a veritable snowstorm. 'Cleaning' an image might take anything from a few minutes to an hour or more, and eliminating dense patches could lead to a significant loss of detail. This process became much easier in 2011, with Adobe's introduction of the 'content-aware' spot healing brush in Photoshop: this could magically, as it seemed to a non-technical user, 'sense' the colours around the blemish and eliminate it without a trace.

A significant part of the cost of many of the specimens I bought arose from polishing so I started to buy via the Web what jewellers call 'rough'. This was generally scanned after coating with Johnson's Baby Oil – a technique used by geologists when working with a microscope – but it was both messy and unreliable: with a scanner the oil was in contact with the glass bed and variations in the thickness of the coating of oil would produce all too noticeable light and dark bands that could not be easily eliminated digitally.

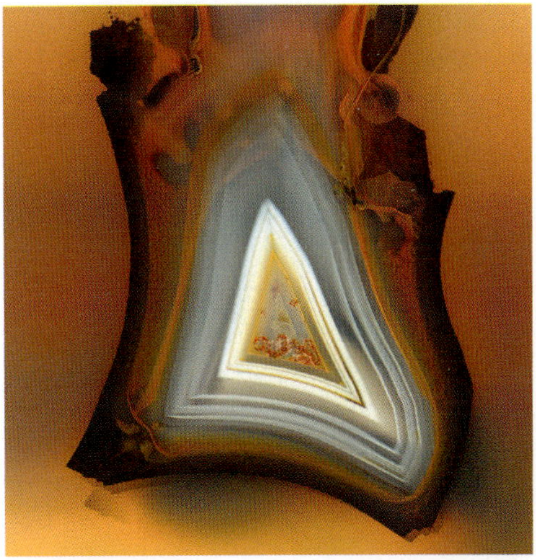
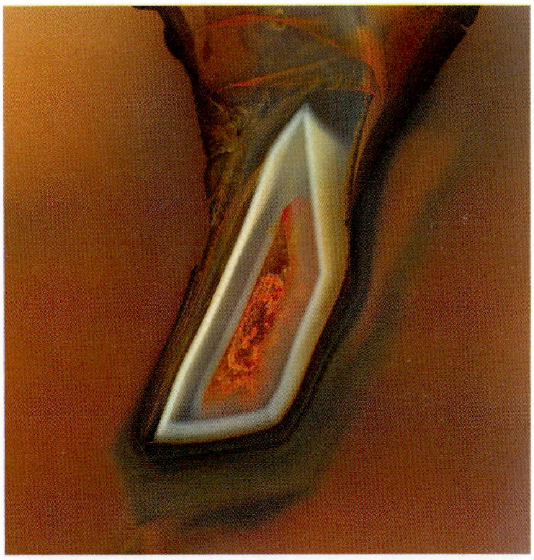
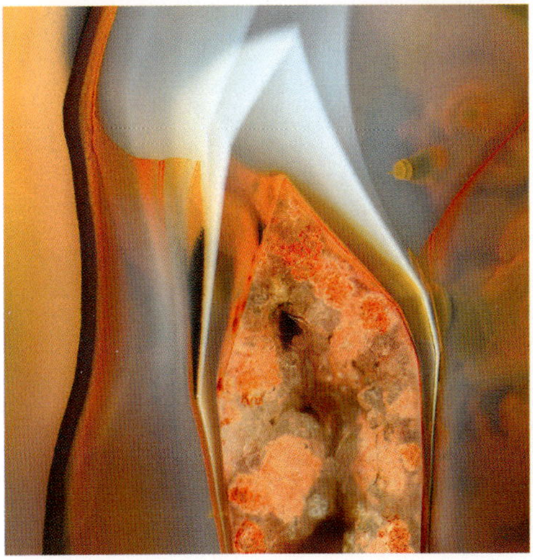
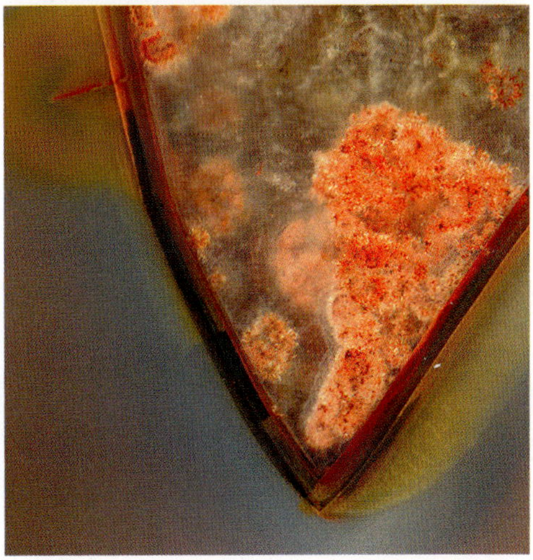
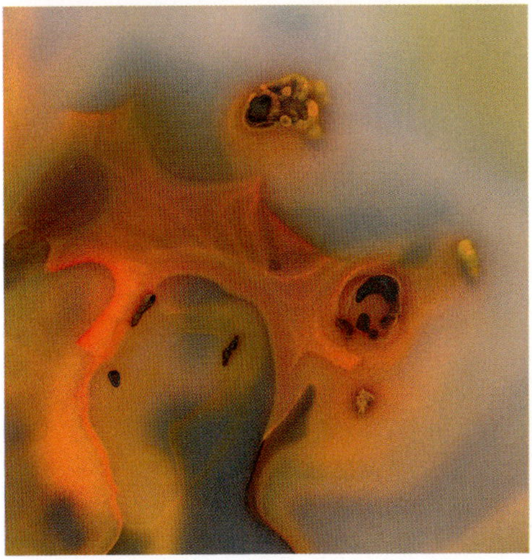

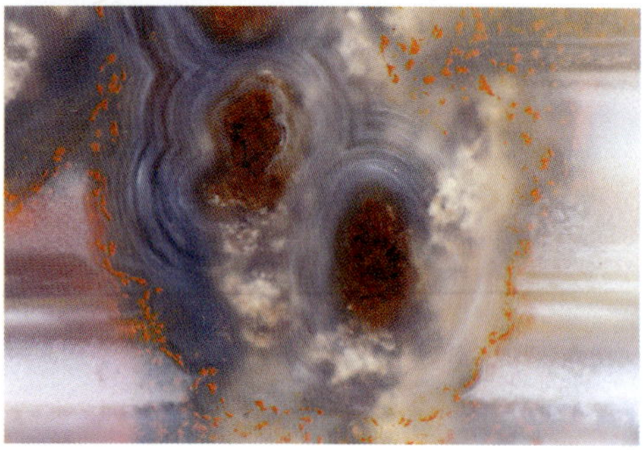

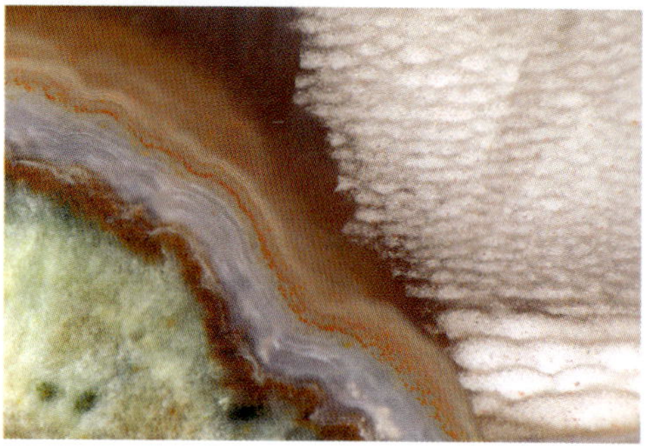

When you hold a slice of a translucent mineral like agate to the light it frequently presents radically different colours to those you see primarily by reflection (as captured by a 'normal' scanner). I realised that to explore such minerals fully I needed a scanner that could operate with both reflected and transmitted light. The principal application of these was to scan photographic negatives or transparencies, and they came with correspondingly higher resolutions, 4800dpi and then, Epson's optical maximum, 6400dpi. This gave an effective magnification of 20x assuming a print resolution of 300dpi, enabling potentially very large prints to be made. But even with that magnification and print size it was obvious that much of interest lay unresolved or completely out of sight. And so, after taking early retirement in 2013, I decided to 'invest' part of my pension lump sum in a microscope. Despite a cost similar to a decent car, I had no regrets: the results were stunning.

I decided on a Nikon system with a 20-200x zoom lens, a purpose-made digital camera and two key devices made by a long-established British company, Prior of Cambridge. The first of these is called a moving stage: driven by two stepper-motors, it can not only position a specimen precisely (to an accuracy of less than a micron) to 'frame' the image, but also be programmed to create large – *very* large – images from multiple frames, which the software automatically fuses, eliminating the joints and differences of lighting. The other device is a motorised focus drive. This allows precise focussing, but also – much like the 'macro' software used with digital cameras – a 'vertical stack' of images can be taken at a precise interval apart (which can be as little as a fraction of a micron). The software then processes these frames – in effect, a digital contour model – to create an in-focus image from which a Virtual Reality 'world' can also be generated by overlaying a 'texture map' onto a three-dimensional surface defined by the multiple triangles of the kind of vector files used to represent terrains in 3D CAD models.

As the illustrations below show, these virtual worlds are unmistakably 'digital artefacts' when compared with the other data from nature, but they have – as I describe later in the discussion of Molly's World – considerable potential for work with children. Matched with music composed in response to their varying 'terrains' they might also make intriguing landscapes of the mind through which to wander. Sadly, the cost of rendering them as full VR environments is, at least for now, prohibitive.

With translucent or transparent specimens, the microscope also allows me to combine transmitted light from below with incident/reflected light from above. The latter is provided by two rings of LED's that can be switched on and off in different combinations, and also rotated: in some

images this can feel like making the Sun come out! Far more than with the scanners, working with the microscope is analogous to taking photographs, and this is obvious in the results: the flat, slightly 'clinical' quality of some of the scanned images gives way to something warmer and – dare I say it? – more 'human' and vibrant.

Before letting these 'Data from Nature' speak for themselves, a few reflections may be of interest. Working away in the evenings or at weekends I came to feel what I can only describe as a sense of duty to present these 'revelations of nature' as directly and as 'honestly' as possible. I would ponder whether an unsightly mark was an artefact of the polishing or scanning process (I have two cats…), or part of the specimen that ought to be preserved. And because so many of the images seemed to me both singular and so characterful I gave many of them names. When friends asked me why I was spending so much time with this work I would say that the best of the images felt like insights into the hidden order of nature. I hope this doesn't sound pretentious. For the first two years I had no thoughts at all about 'applications' and the feeling of 'cosmic connection' some induced is still vivid. It is born, in part, of an almost lifelong fascination with forms and patterns in nature. As a student of architecture I was introduced to the 'Vision and Value' series of books edited by Gyorgy Kepes at Harvard. Featuring essays and images by scientists, engineers, architects, designers, composers and artists, they were published in the hope of finding visual languages that could provide common ground for the divided worlds of art and science.

My colleagues in the school of architecture did not share my fascination and several were particularly uneasy about the images that evoke 'real' landscapes or other subjects. With level-bedded agates, below, often called 'sea and sky' agates, this is hard to avoid: even those without obvious 'clouds' or 'waves' are difficult not to see as more or less abstract seascapes. Other images function as what psychologists call pareidolia, a mental

phenomenon in which the mind responds to an image by perceiving a familiar pattern where none exists. Polonius' 'camel in the clouds' in *Hamlet* is the most famous literary example, but the one I like best is in a 'Peanuts' cartoon. Lucy, Charlie Brown and Linus are lying on a mound gazing up at the sky. 'If you use your imagination, you can see lots of things in the cloud formations,' announces Lucy: 'What do you think you see, Linus?' Clever-clogs Linus's imagination is predictably vivid: the map of the British Honduras; the profile of Thomas Eakins, the famous painter; and the stoning of Stephen – 'I can see the apostle Paul standing there to one side' he adds by way of confirmation. There is a serious point here: once our brain starts running with a particular 'reading', any amount of detail will be incorporated or ignored to make it convincing. Several of the images that follow provoke such responses – children especially delight in letting their imaginations loose in inventing interpretations – and this 'pareidolic' quality was certainly why I captured some of them.

But as I think about this I am less and less sure that many of the images are pareidolia in the strict sense. This is especially so in the case of the 'work' of Carlo Alcite, for which I have reserved a chapter of its own, but it is also true, I think, for many others in which the processes of formation are *analogous* to those at work in the larger-scale world we inhabit, and hence, by sharing in the subtle variability of nature are intensely adept

at evoking it. The rock strata of a real landscape are layered and deformed in ways that can find precise analogies in the space of a millimetre or two of a mineral. Far from troubling, I find this thrilling, and will return to it in the conclusion.

I will end this introduction to the images by relating a story of what might have been. Some ten years ago the National Museum of Wales had an unexpected opportunity for a small exhibition in new accommodation that was going to be ready ahead of schedule. Knowing about my work – with, by then, its early textile and ceramic prints, and a collection of 'Frocks from Rocks' – the Keeper of Geology thought it could be the basis for a fascinating show of 'work in progress' that, uniquely, touched on most of the Museum's departments. He arranged a meeting for me to present a selection of images to the Director, who seemed to find many of them beautiful. My rising hopes, however, were quickly dashed: 'I can't show them because I couldn't position them intellectually,' he said. Asked to explain he added, 'people would mistake them for art.' Quite why photographs of minerals (for that's what these images are, albeit made with devices other than a normal camera) couldn't be 'art' when their camera-made cousins have long been admitted into the hallowed precincts of the gallery eluded me.

Most of the 'Data From Nature' that follow were intended either as evocative images (their success is for the reader to judge) or as 'neutral' records of interesting 'facts' of nature. Increasingly, after acquiring the microscope, I look for details that capture – or appear to capture – the processes at work in their formation, and in doing so seem to evoke the vast scale of the geological processes described in Chapter 1. It needs only a millimetre of an agate to suggest cataclysmic events, and I cannot help marvelling at this. Surprisingly few of these images have been made with particular applications in mind: for the most part the scarves, dress fabrics, materials for interior design, and 'digital paints' discussed later were drawn from the ever-growing database.

The images that follow are grouped by mineral type, with short texts or captions providing factual information, in so far as I can offer it, about their formation. But as I explained in Chapter 1, in many cases there is little that can be said with any degree of certainty: of that which I cannot speak, I have remained silent.

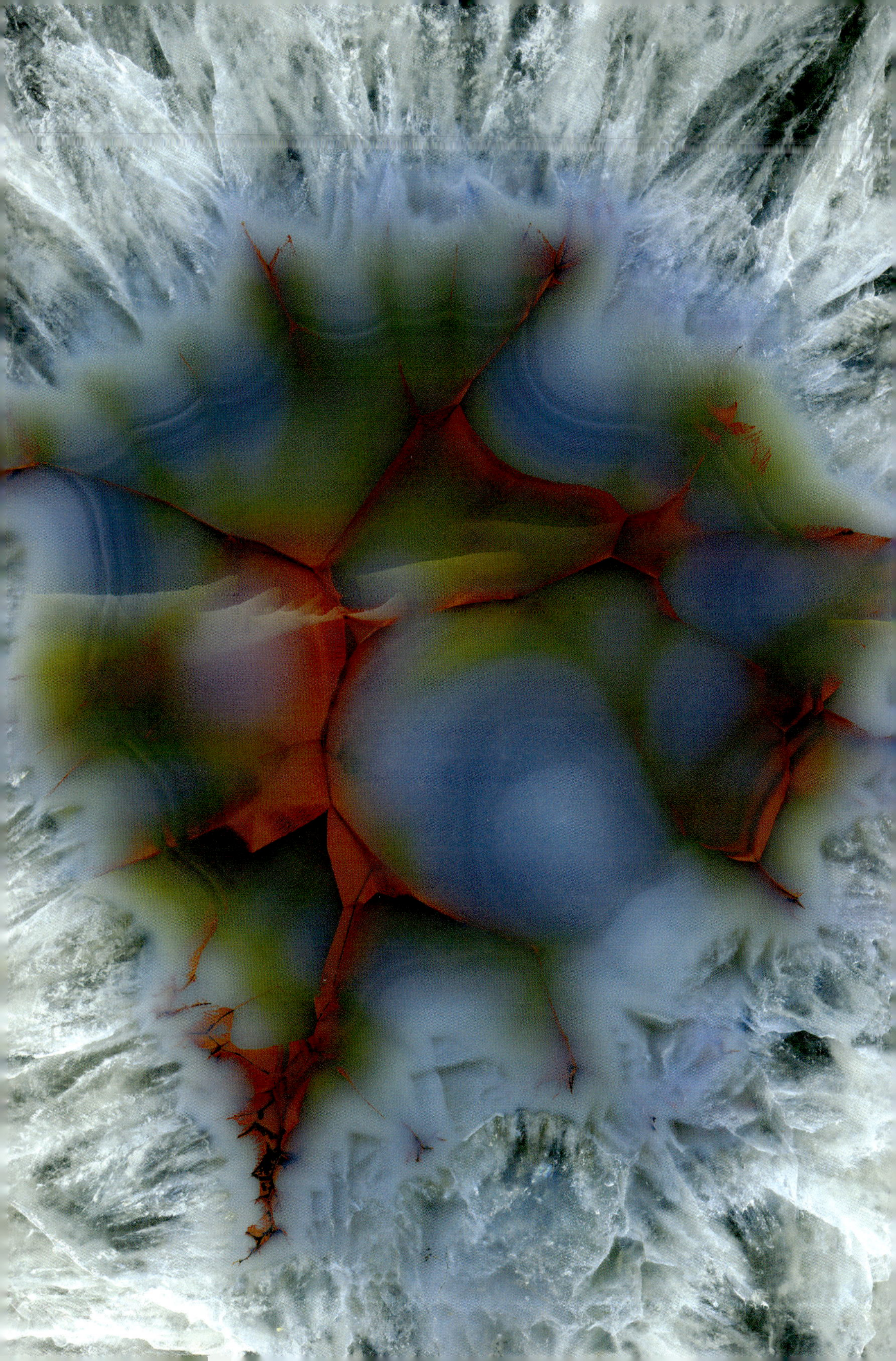

AGATES

Mineralogically speaking agates are a banded, microcrystalline form of chalcedony, which in turn is a kind of quartz. They typically form in potato-shaped voids in volcanic rocks like basalt, and being much harder than even their durable hosts eventually weather out as nodules which collectors saw open and polish. Now, with a global trade at their disposal, large-scale suppliers mine for them in productive rock formations. Specimens more than a few centimetres across generally reach the market as slices a few millimetres thick.

There is no agreement among experts as to how agates form, although in highly magnified images (e.g. that on p.52, bottom left) you can clearly see linear crystals growing outwards from a centre to form a spherical shell. The diagrams suggest how such growing shells might intersect to form the familiar large-scale banding patterns, but there is no certainty as to how this actually happens. In looking at pictures of agates one must always remember that the surface we admire is, like an architect's plan, a cut through a three-dimensional form. The image on p.52 is at the limits – 200x – of my microscope and greater magnification still would reveal that the linear crystals do not, like those of so many minerals, grow in straight lines but as elongated helices, like a strand of DNA.

There is often insufficient agate-forming silica-gel to fill the host void, and in these cases the centre is commonly occupied by clear or purple (amethyst) quartz. More unusually, as in the example shown here, an 'island' of agate can form in a 'sea' of quartz. In some cases, however, it appears that new agate-forming material later enters the cavity and fills the void with horizontal 'level-bedded' bands: these form under the influence of gravity, not by growing out from the host rock, and are the source of the 'sea and sky' images mentioned above.

The majority of agates on the market come from the world's largest supplier, Brazil, and are dull brown or grey in colour. These are often artificially coloured to sell cheaply world-wide and, sadly, are frequently found in shops in natural history museums which really ought to know better than to pass them off as 'natural': their garish neon pinks, blues and greens are a parody of true agate colours whose infinite subtlety, like their overall formation, is far from fully understood. The colours we see are a result of both natural pigmentation and the movement of light internally: all agates are to some degree translucent, so light can be reflected and absorbed in complex ways. Hold a slice up to the light and many will appear as milky creams and varied browns (see p.53); seen in reflected light, however, the same specimen (opposite) yields brilliant reds and blues. Agate bands are highly susceptible to the presence of elements that may enter during the process of formation: it is thought that as little as one part in a million is sufficient to change the colour dramatically. Substantial inclusions of other matter can not only produce variations in the colours but also lead to the formation of what are known as moss agates, with their characteristic vegetation-like growths. The combination of manganese and iron oxides may also lead to the growth of branching dendrites like those found between the bedding planes of limestone (p.221). Dendritic agates are particularly common in India, and highly prized by both collectors and jewellers.

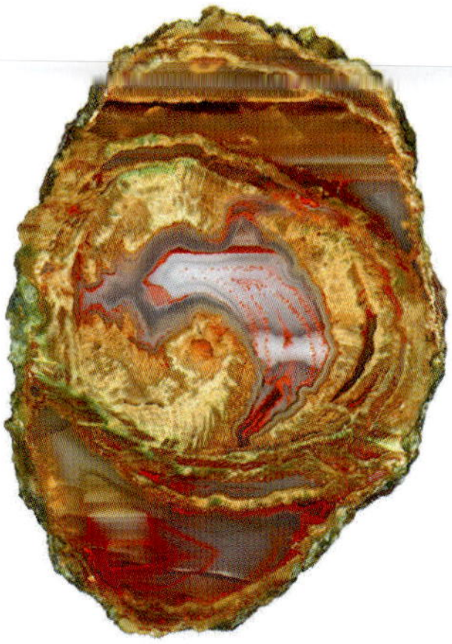

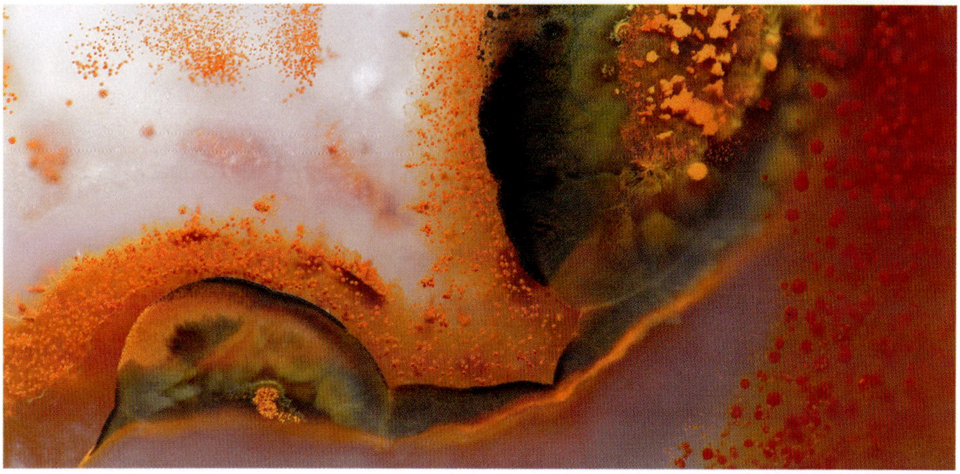

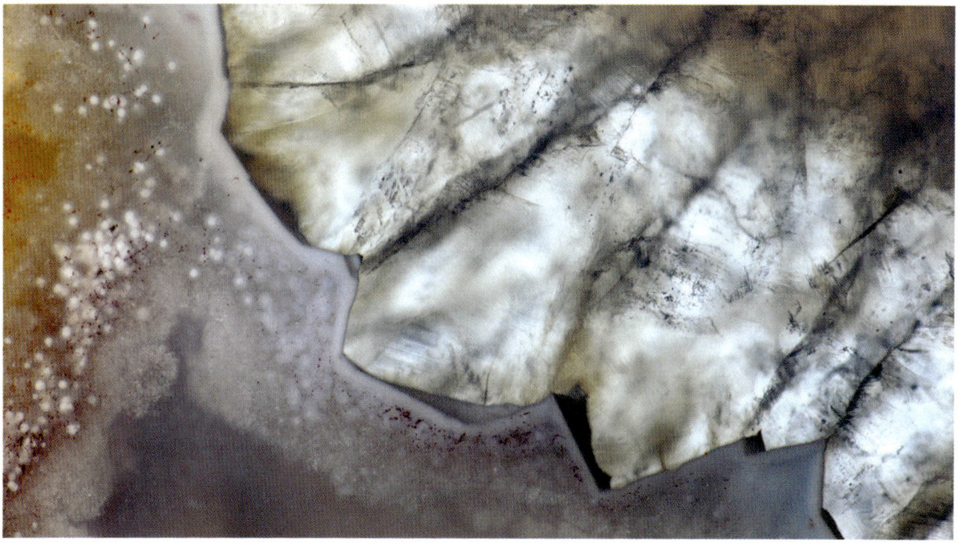

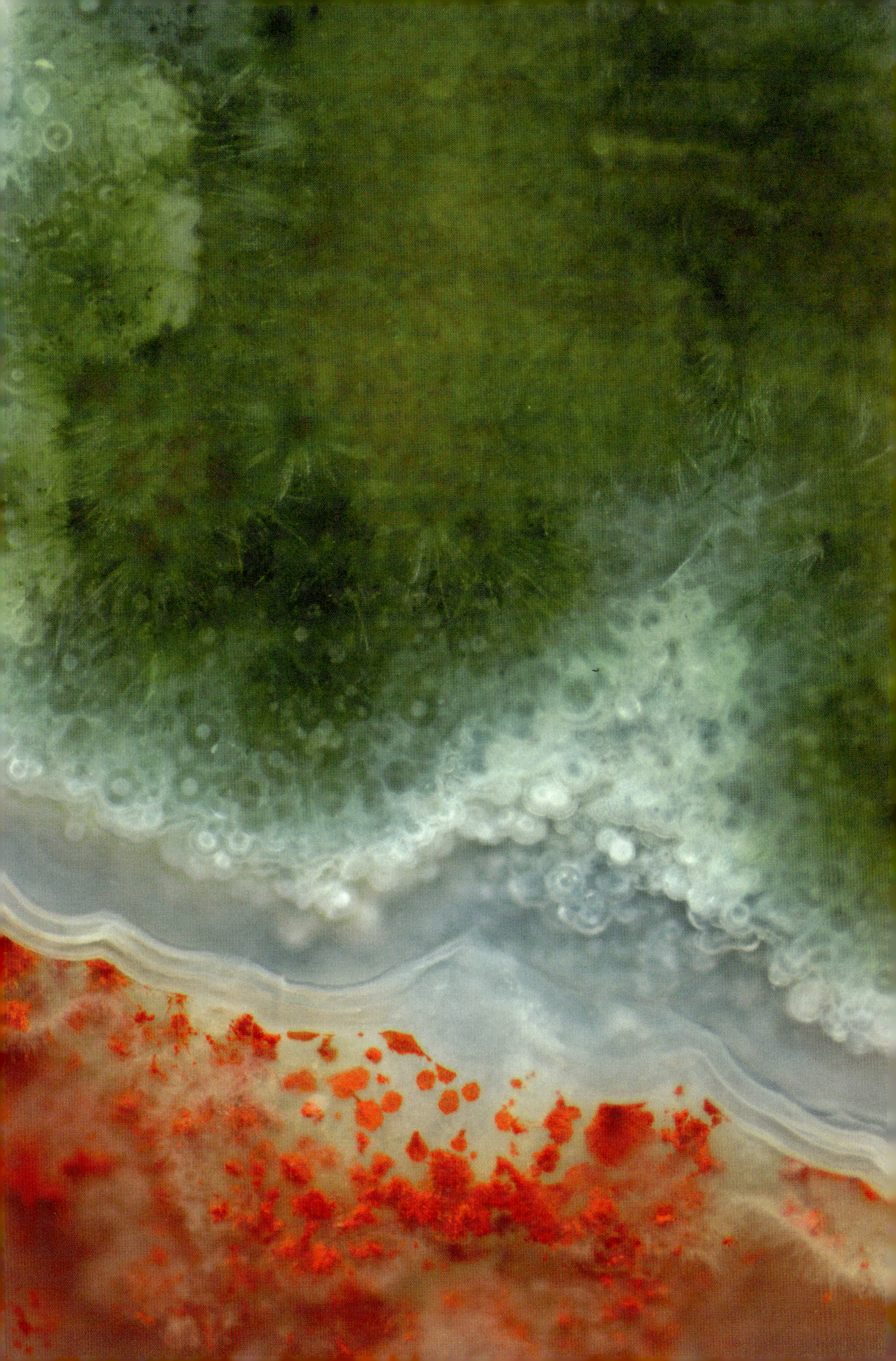

QUARTZ

Quartz – silicon dioxide – is the second most widely occurring mineral at the Earth's surface, constituting some 12% of its crust. Its purest form, optically clear rock crystal, usually occurs in crevices of metamorphic rocks called schists, but may also be found in volcanic (igneous) rocks. The solution from which quartz crystallises frequently contains small amounts of other minerals, as well as bubbles of gas and water, that cannot be accommodated in the crystal matrix. Captured as mysterious 'inclusions', these may include needles of actinolite, tourmaline or rutile; manganese dendrites; flakes of green chlorite; spots of blood-red haematite; and particles of clay to which other elements may adhere.

The normally clear quartz crystals may also be coloured by small traces of other minerals and elements. The most familiar, thanks to its frequent occurrence in geodes, is amethyst, which owes its red-violet colour to radioactive irradiation and an admixture of iron. The pale yellow of the far rarer citrines is caused by iron trioxide, while the grey-brown colouration of so-called smoky quartzes is caused by silicon ions being released from their positions in the crystal structure during radioactive decay. Since the ancient world, craftsmen and jewellers have been aware that these naturally occurring colours can readily be manipulated by heat and radioactivity: heated to 250°C, for example, amethysts take on the colour of more valuable citrines.

Quartz is among the most stable and durable of naturally occurring materials, but when subject to powerful physical stresses – as frequently happens during the aeons that may pass before it is uncovered – tiny fractures can result, either of the conchoidal (shell-like) form characteristic of homogeneous materials, or along the otherwise invisible planes of the crystal lattice. Like the inclusions of other materials, these fractures may reflect rather than transmit light, appearing from certain angles as miraculously thin mirrored surfaces.

Large crystals may take tens of thousands of years to form but then, subject to the exigencies of later fractures, the intricate formations they contain are literally frozen in time. This long process of formation is, however, belied by the crystals' appearance and the almost snapshot-like images they yield. In the miniature worlds captured from quartz, streams of bubbles appear to rise towards unseen surfaces and diaphanous creatures float in strangely coloured seas. The most intriguing among the potential subjects for the microscope are called secondary liquid crystals (pp.64-5). When a crystal cracks internally, liquid silica may be drawn in by capillary attraction. Like a perfect glue, this can 'heal' the fracture completely, leaving no trace. More commonly, it is thought, the repair is not perfect and tiny secondary liquid crystals are left behind, each containing a diagnostic bubble of gas. These may appear like shoals of almost identical crystals ranged across the plane of the crack, thousands upon thousands of them, each a few microns across. Or some may be much larger, up to a millimetre or more long and evoking by turns strange fishes with gas-bubble eyes, oddly shaped cutlery or mysterious surgical instruments.

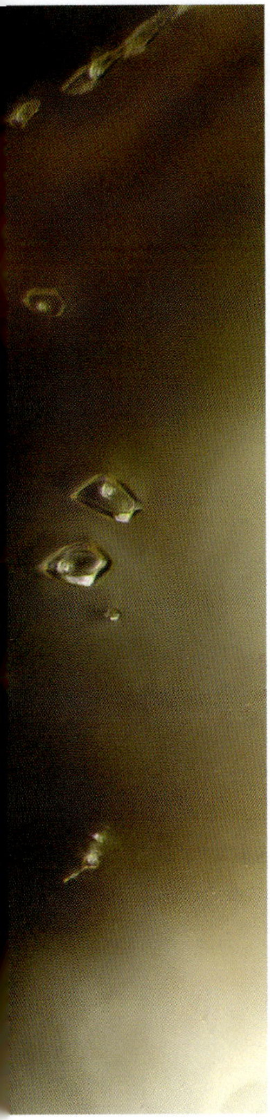

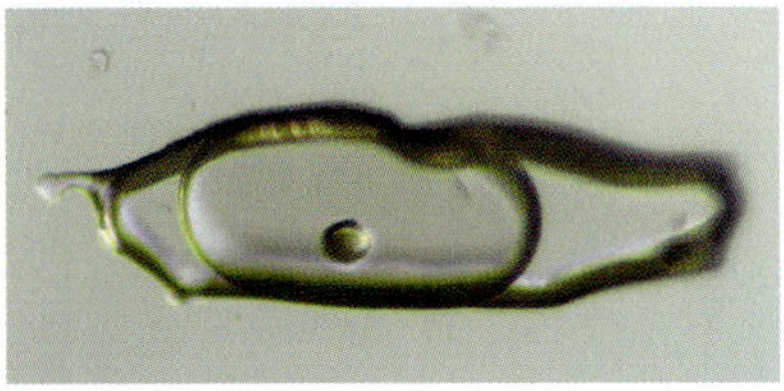

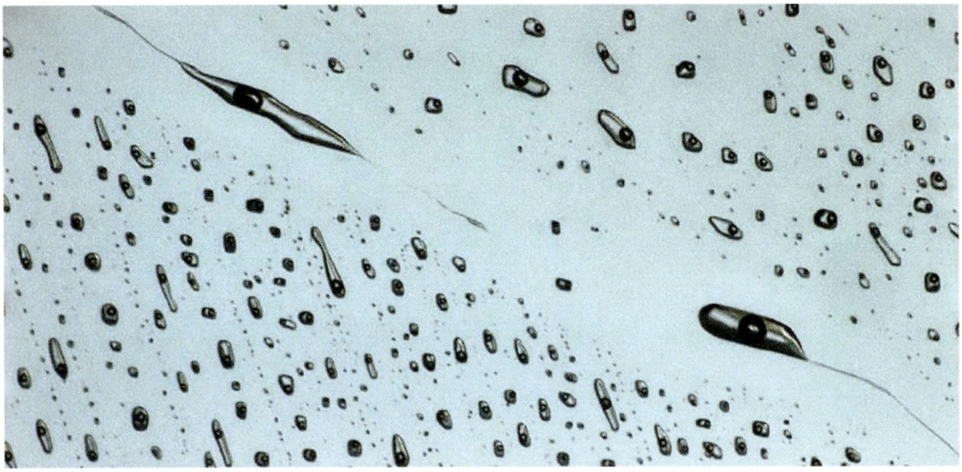

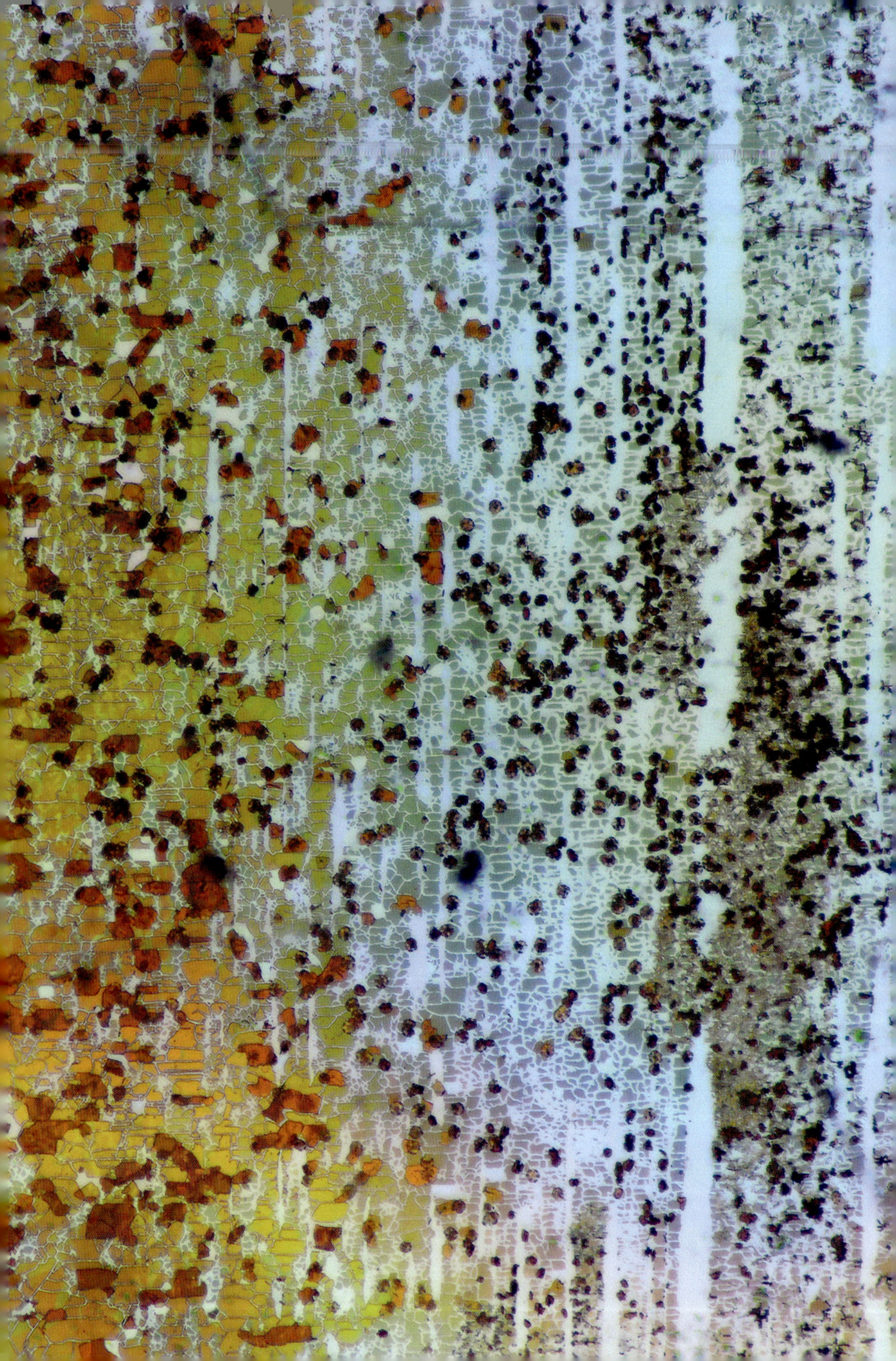

MICA

Mica is found world-wide in all the rock types. It forms in vanishingly thin layers created by the hexagonal, sheet-like arrangement of its atoms. The largest documented single crystal of mica was found in Canada and weighed 330 tonnes. I have one or two solid, opaque specimens but the samples I have worked with are a dozen or so layers thick, like a piece of stiff paper, and a few centimetres wide. I found them, mostly in a sorry, desiccated state in a drawer in a leading London mineral dealer, Gregory, Botley and Lloyd (founded in the mid-nineteenth century but sadly no longer in business): they had inherited them from the previous owner and had no idea why they were there, mica in this form being of no interest to collectors.

What interested me were the fragmentary traceries of dark grey and red/yellow streaks of colour that reveal the crystal lattice. Made principally, I guessed, by the usual culprits of manganese and iron, they yielded pleasingly decorative designs to the scanner and, later, a veritable wonderland of bizarre formations to the microscope, including a repertoire of child-pleasing 'creatures'. The presence of these delights – as close as minerals get to visual humour – was due to the delamination of the mica, probably due to relief of pressure in the environment of formation, allowing the water-borne elements to be drawn in by capillary attraction. I have no idea if they were the result of arrested growth or microscopic erosion, but like many of the images in this chapter they seemed to me vivid illustrations of a remark by Kepler in his essay on snowflakes: 'Formative reason does not act only for a purpose, but also to adorn. It does not strive to fashion only natural bodies, but is in the habit also of playing with the passing moment, as is shown by many minerals from mines.'

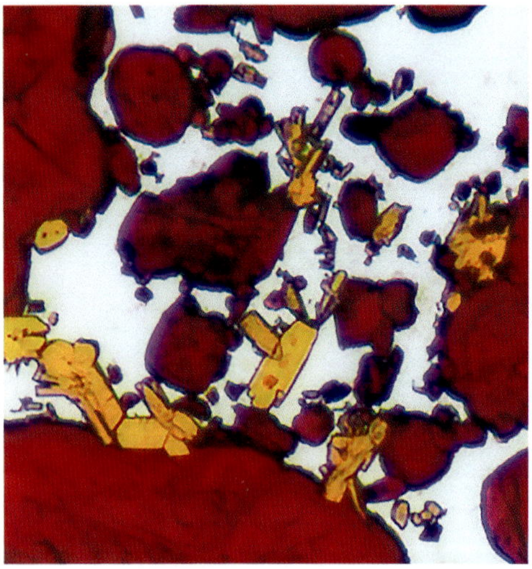

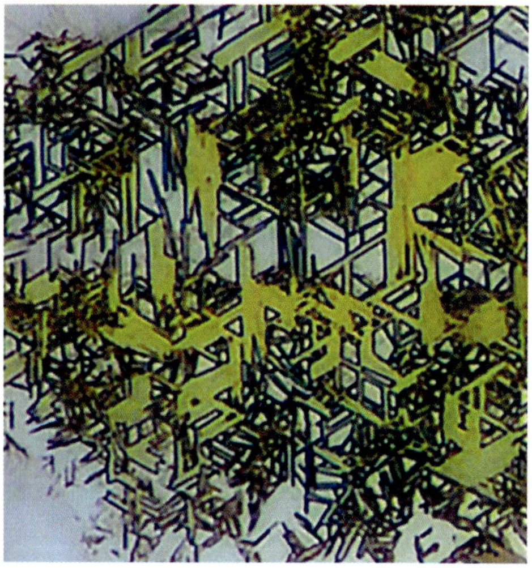

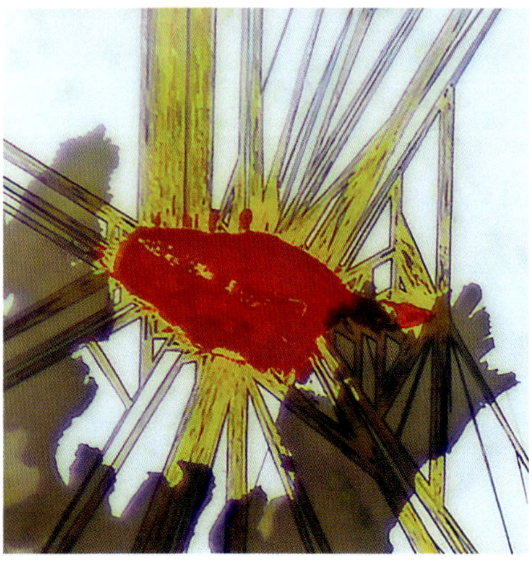

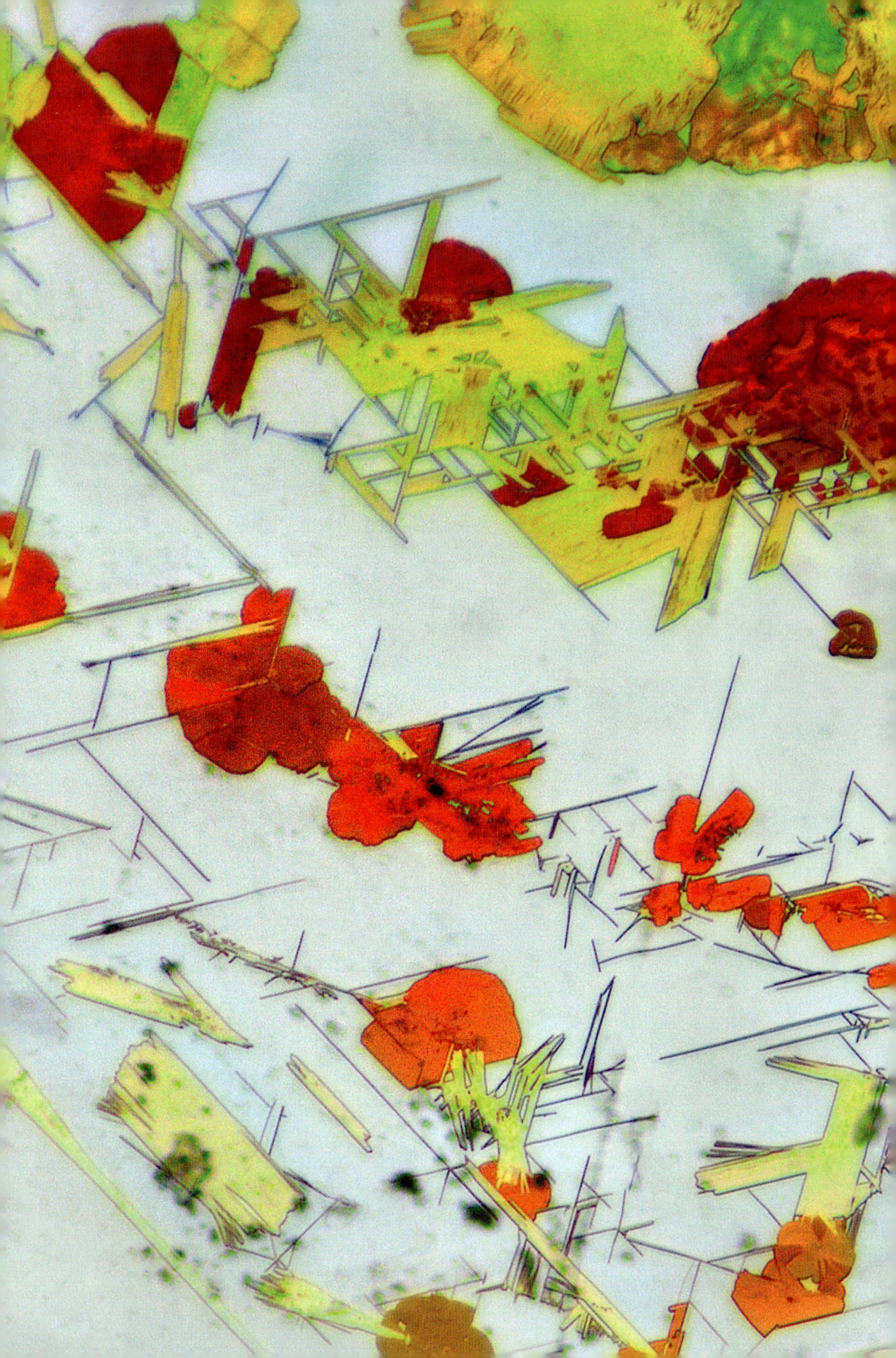

FLUORITE

Fluorite has been called 'the most colourful mineral in the world.' It is readily tinted by elemental impurities, as well as exposure to radiation, and every color of the rainbow can be found in fluorite specimens. The dominant colours commonly form zigzag bands, which are then overlaid by diaphanous drifts of colour created by chemical impurities. To the eye, and at low magnification, the typical effect is 'atmospheric', but seen through a microscope an elemental architecture appears, based on the cubic crystal structure.

The name fluorite is derived from the Latin for 'flow' and it has long been used as a flux in smelting iron. In 1852 fluorite gave its name to fluorescence, a natural phenomenon caused, in fluorites, by impurities found in a few locations.

JASPER

Like agate, jasper is a form of micro-granular quartz and chalcedony, but unlike agate it is opaque rather than translucent. When I started scanning, the most commonly available was a type of orbicular (named after the many circular/spherical formations) jasper: the orbs are thought to form around an 'irritant' particle and to consist of spherical growths of linear quartz and feldspar crystals. Known commercially as Ocean Jasper, this material was to be found in every shop I visited and I assumed it must be rather common. In point of fact, the name turned out to be a trademark of an American dealer and the material was from one remote location on the north-east coast of Madagascar, where it was only visible at low tide. It had been discovered in the early 1900s but the location was lost and only re-discovered in the 1990s. In the year 2000, Ocean Jasper made its public appearance to great acclaim at the world's largest mineral show, held each January in Tucson, Arizona. Sadly, this fascinating material is now mined out and no similar deposits have been found.

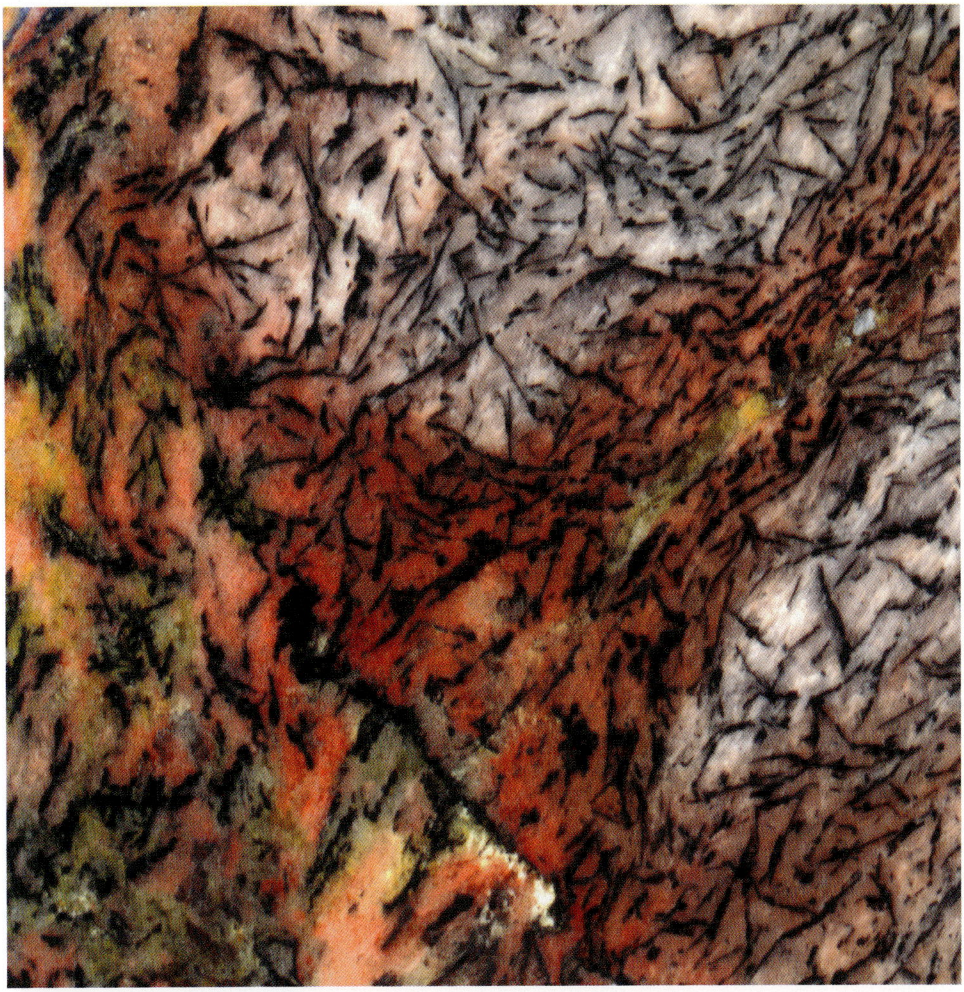

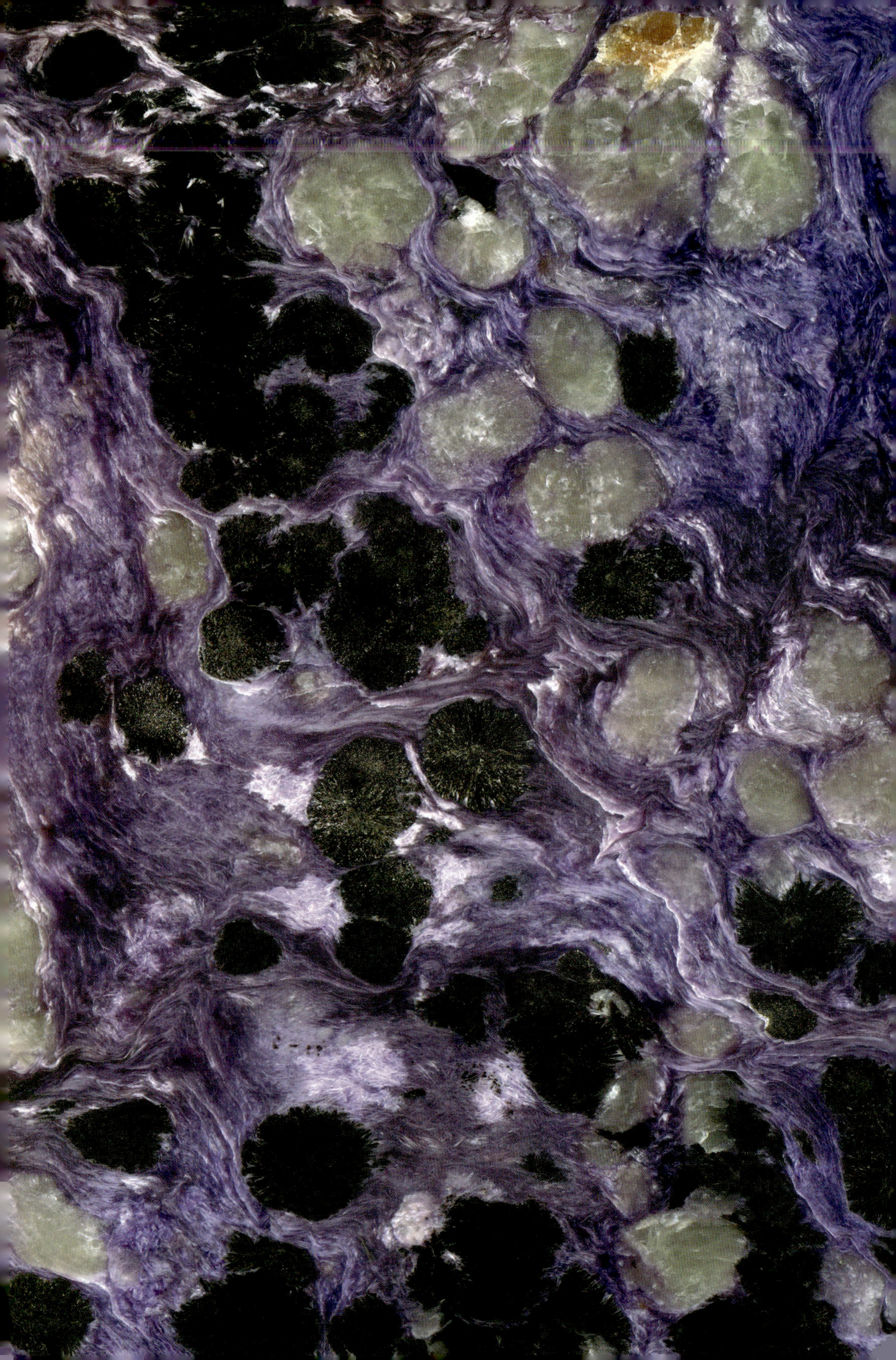

DIVERSE MINERALS + OTHER MATERIALS

I have captured data from many other minerals and some of my favourite images are shown on this and the following two spreads. Purple-coloured streams may not be found in large-scale nature, but the Russian mineral charoite – shown here only a little larger than life-size – offers extraordinary verisimilitude to the eye and, as seen on the following page, when magnified over 100x.
Few, if any, man-made materials can match the fine grain of many minerals, but I have worked with several that yield fascinating results. I refer to them as 'honorary nature' and examples from ceramic glazes, glass, Fordite and squeezed paints ('Strizzate') follow the mineral selection.

Opal in rock matrix

Hypersthene

Tourmaline

Meteorite

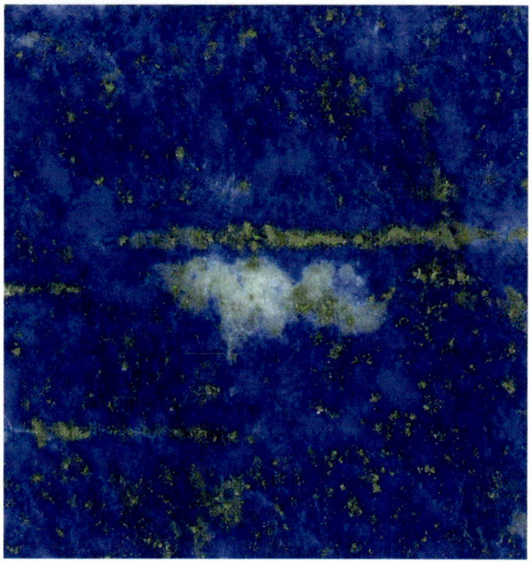

Lapis lazuli

Paesina stone

Charoite

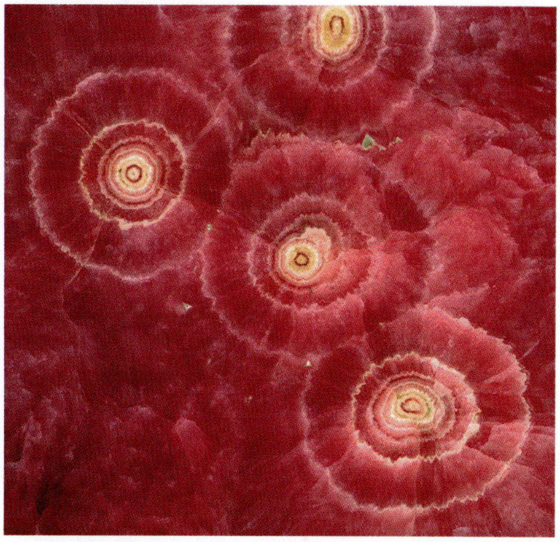

Rhodochrosite

Pyrite on azurite

Malachite

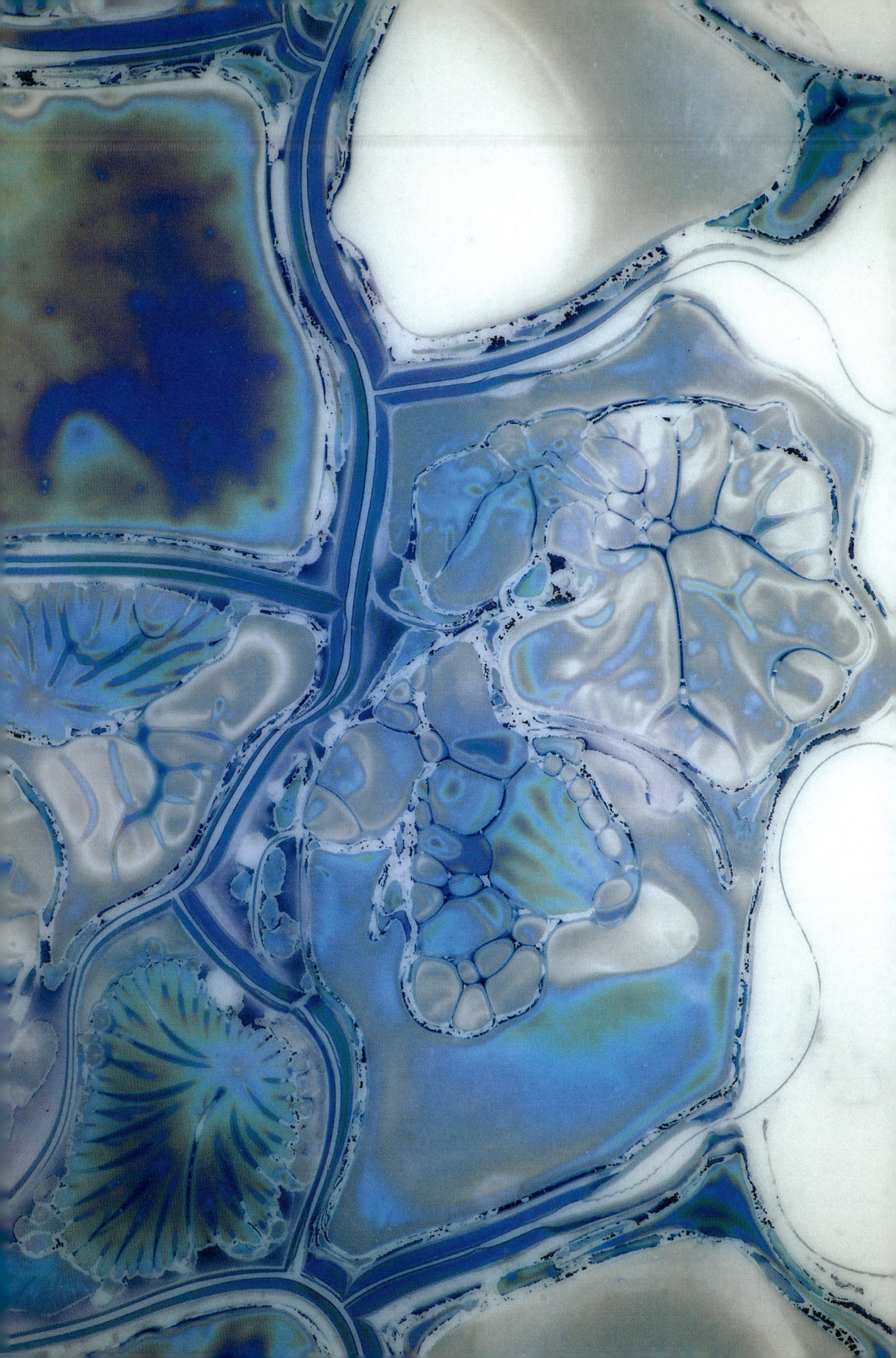

CERAMIC GLAZES

The images in this section come from two sources: small dishes by Simon Rich, a potter based in West Wales, and flat tiles specially prepared by Geoff Swindell, who is renowned for his innovative glazes. Rich's crystalline glazes use a process developed in Japan that involves throwing salt crystals into the kiln during firing, following which they undergo rapid growth creating patterns like the blue, flower-like forms on p.86 and the pulsating 'waves' and 'rocks' shown on pp.90-91. When I say that Swindell's tiles were 'specially prepared' I mean only that he normally works on characterful forms inspired by marine organisms, not anything so prosaic as flat tiles – which, of course, are better for 'digital scrutiny'. The glazes have not been manipulated to give particular results: the patterns formed during firing from the interaction of the constituent elements, and such unpredictable, 'bottom up' formation is key to all these examples of 'honorary' nature.

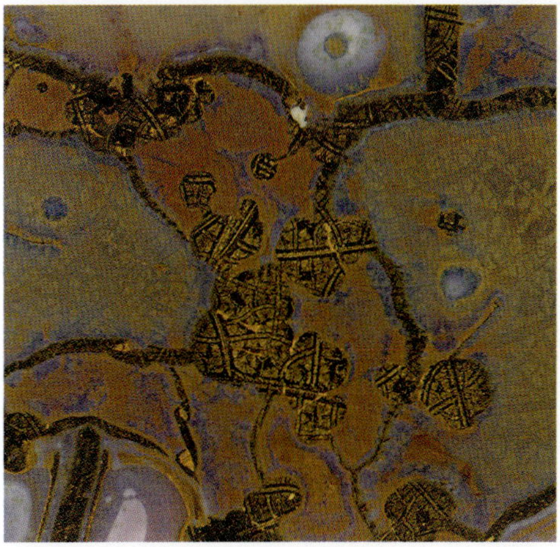

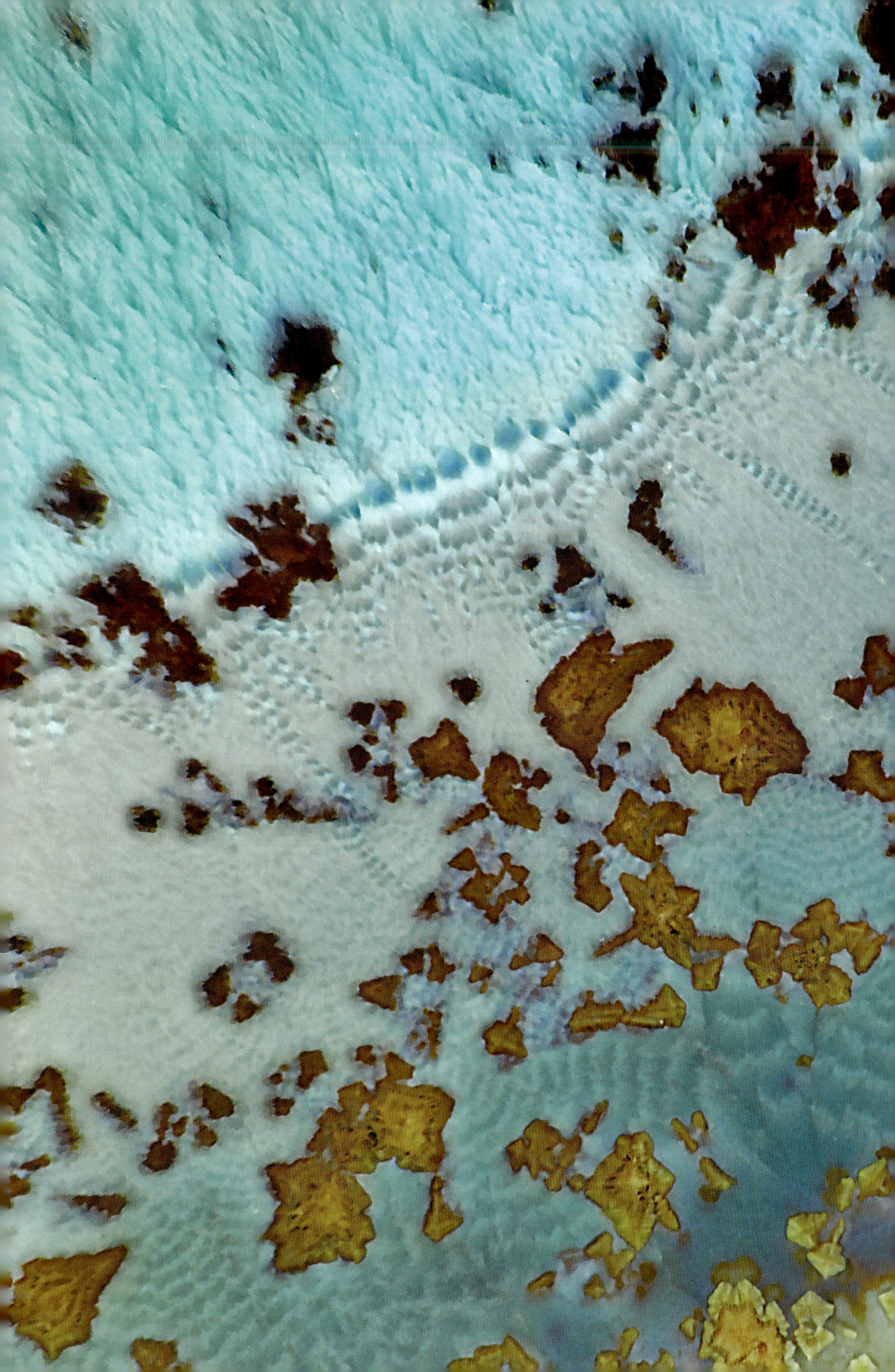

GLASS

Made primarily from silica sand, glass shares some of its properties with quartz. It is not as optically clear as the finest quartzes, but it has a granular, homogeneous texture that also breaks into conchoidal fractures, named after similar, unrelated patterns found on mussel shells. The image opposite shows the edge of a 1.5mm thick laboratory slide. Along its right-hand edge you can see a thick, gritty line where the glass-cutter scored the surface before breaking. The conchoidal fractures to the left would then immediately have propagated along its length at lightning speed.

The other images are all from induced conchoidal fractures. The cliffs that evoke a 'coastline' are tiny and created by rolling a screwdriver shaft along the edge of a glass slide. The next were made by 'knapping' a thicker piece of glass with a section of deer antler in much the same way that our distant ancestors worked flint. The images on pp.93-95 were produced by an altogether more obscure process about which even Google is silent. Known as glue-plucking, it is thought to have been discovered following the 1845 repeal of the glass tax in the UK by Sir Robert Peel's government. As large shop windows came into vogue, posters were glued to them and when eventually peeled off they 'plucked' part of the glass surface. The effect was decorative and became part of the glass-worker's repertoire. It involves roughening the glass surface – sand-blasting is the preferred method – and then applying a coating of a glue derived from the stomach of a cow. The glue binds to the surface and the collagen fibres in it shrink. After a few days, in the right conditions of temperature and humidity, the glue can be ripped off, pulling fragments of glass with it and leaving behind a textured surface made of a mass of large and tiny conchoidal fractures. Under the microscope these are revealed to combine graceful curves with gloriously odd details.

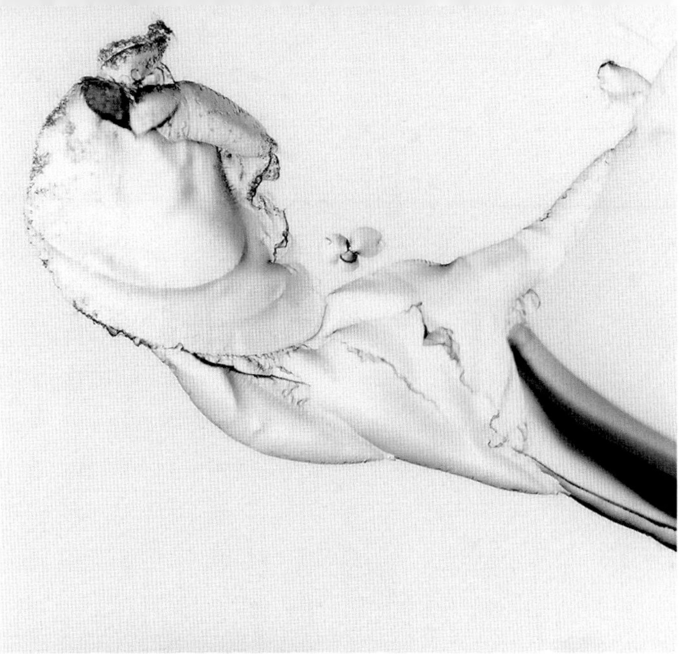

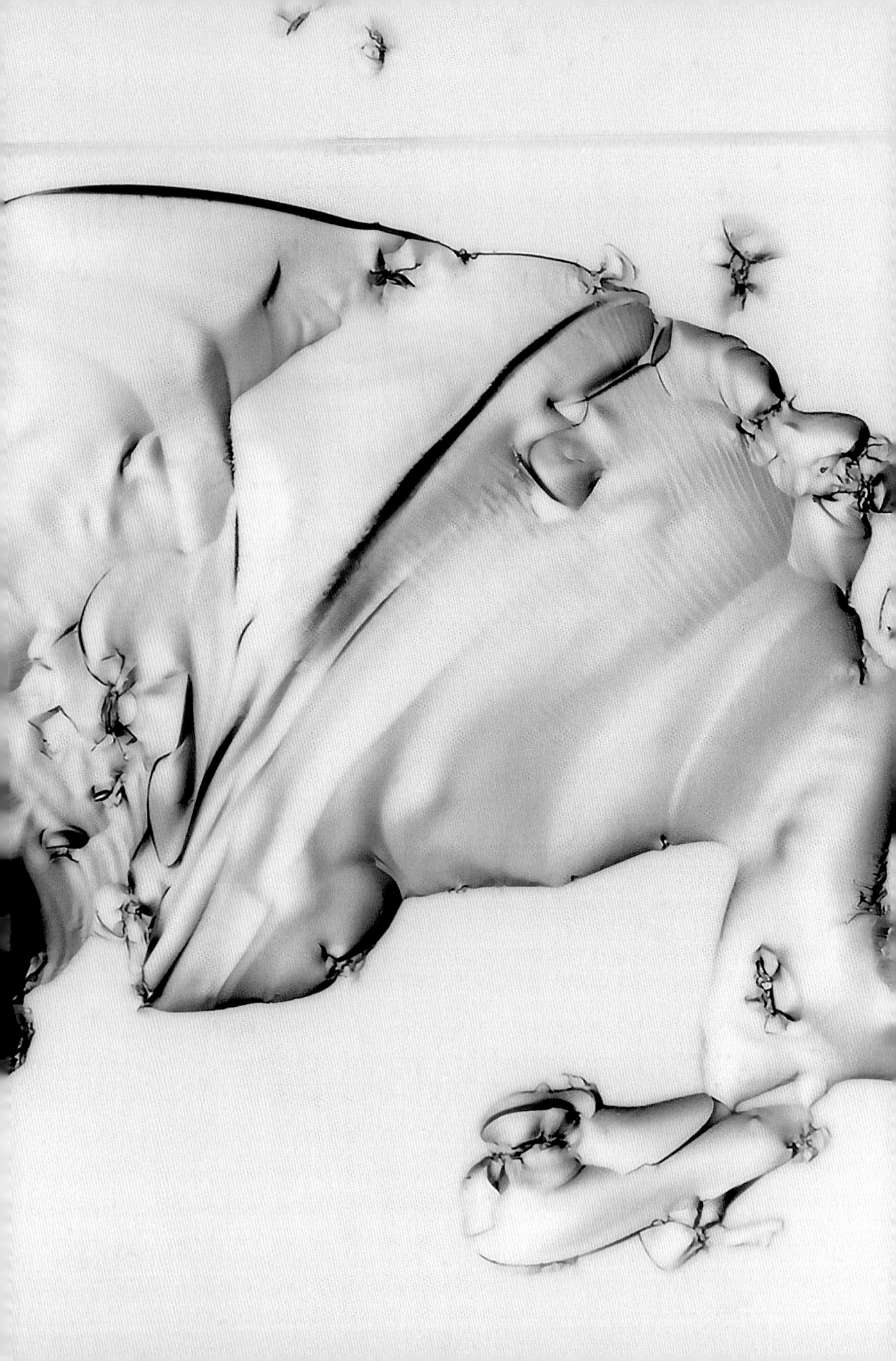

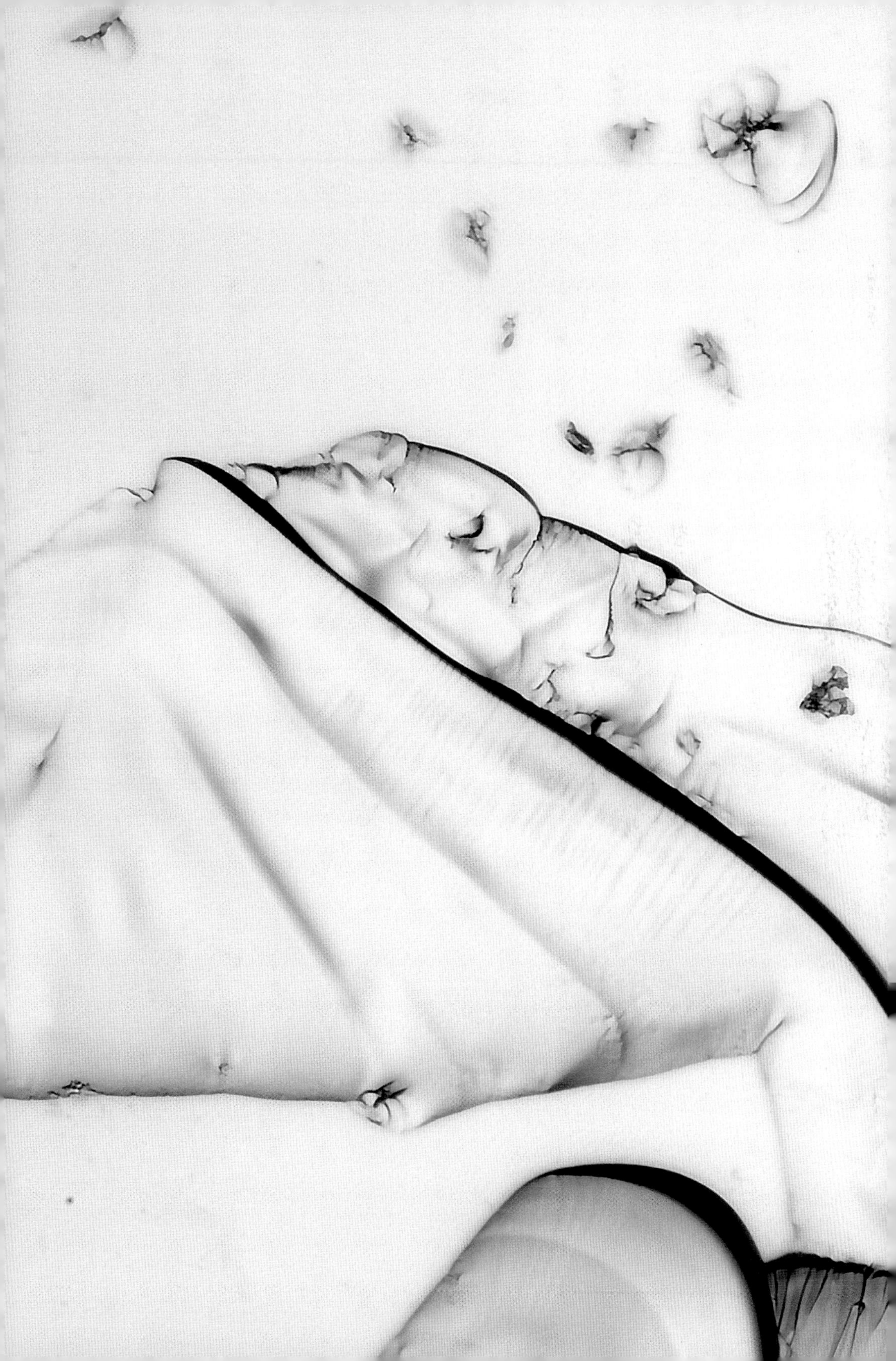

FORDITE

Known as man-made agate, Fordite comes from the spray booths of the Ford car manufacturing plants in Detroit and Dagenham. In the pre-robotic 1970s cars were sprayed in rooms that doubled as curing ovens. Large amounts of paint were deposited on the floor and this built up into a thick layer of coloured 'strata' – the geological term seems apt. When the painting process changed someone had the bright idea of cutting up the coating, which was sufficiently hard to cut and polish. Not surprisingly it soon found a ready market in the jewellery world: the layered colours were intrinsically appealing, and a carefully chosen pendant or brooch allowed the owner to identify the precise colour of their first Cortina or Capri.

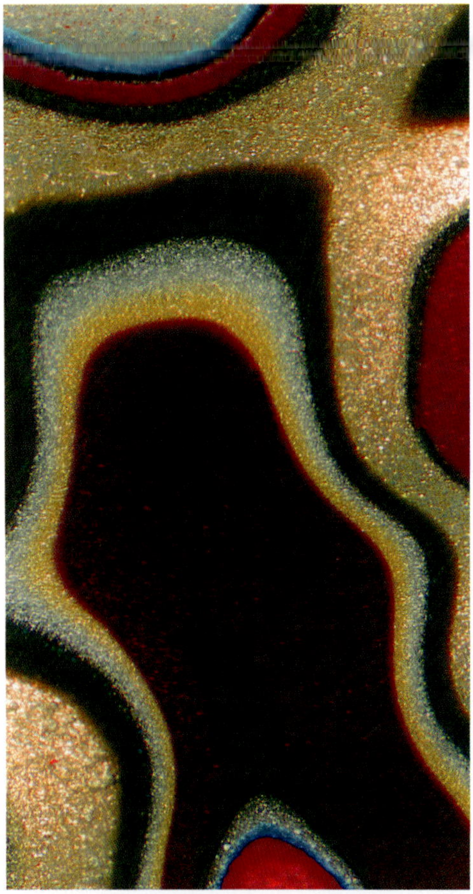

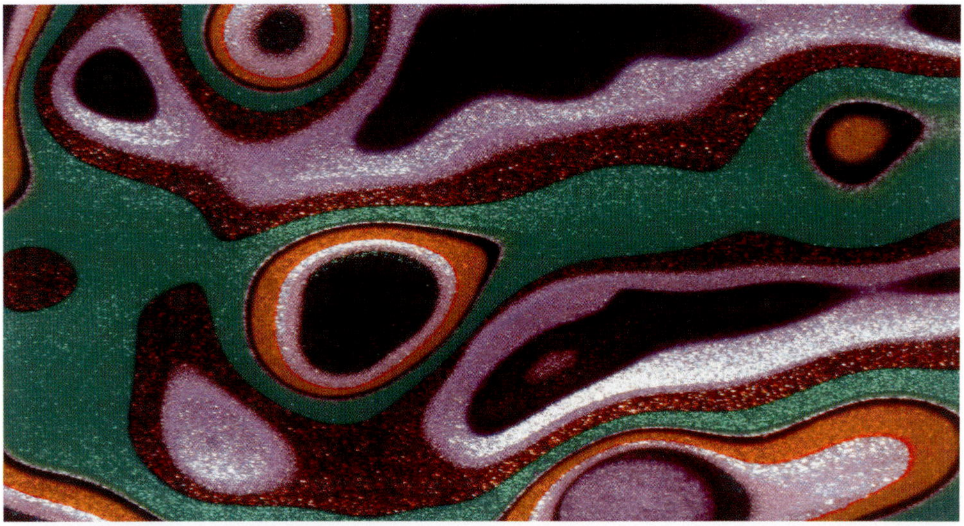

STRIZZATE

During a first-year project on 'Process and Form' one of my students decided to experiment with squeezing paint in a folded sheet of paper – the Rorschach-blot-like 'butterflies' that typically appear are familiar to most children. Noting the fineness of some of the detail I asked to borrow one to scan, and the result was more than promising. Several hundred trials followed using high-quality artists' acrylic paint (oils were insufficiently viscous) and made, to avoid the symmetry, by squeezing and manipulating the paint between two pieces of glossy paper. They worked well and some, slightly to my surprise, packed sufficient variations into a small space to reward scrutiny with the microscope. They eventually found application as a scarf collection, for which the word 'paint squeezes' seemed less than ideal: hence the rather more elegant Italian name, 'Strizzate'.

3

APPLI-
CATIONS

I had been happily scanning away for two years when a friend mentioned that he had been told that digital images could now be printed directly onto fabrics, and suggested that the fashion world might offer a good outlet for my 'data'. It turned out that the technology was surprisingly new: the first producers in Britain – the Silk Bureau – had set up in 2000 and were still working from a double garage (they now have a large, successful factory); the pioneers in Italy, Maver of Como, had switched to digital in 1997 to widespread derision from their silk-screening competitors. The digital facility is located in a climate-controlled red box in the middle of a vast, otherwise largely disused factory, which began life as a place for weaving silk before turning to silk-screening: the data formerly stored on the empty racks of the silk-screen store (opposite) would now fit on a pocket-sized digital drive. Despite its comparatively diminutive size the digital unit houses some very large machines and produces considerably more than used to flow from the silk-screen tables.

Digital printing has two decisive advantages: firstly, it can reproduce the subtle variations that are a hallmark of most of my data; and secondly – as with all forms of digital production – it does not need expensive screens, and hence can produce one-offs and short runs economically. That, at least, was the theory. In practice digital printing was very expensive, and the only retailers who could then afford to sell short-run, digitally printed goods were, as the saying goes, 'high end'.

I sent booklets illustrating my wares to various fashion and interior designers, but none responded. It quickly became clear that a total unknown, and from Wales of all places (many of London's metropolitan elite seem to regard Welsh architects and designers with particular contempt), was unlikely to find an easy route to market. One well-known interior designer assured me he couldn't imagine any *London* designers would be interested. And there was another, practical, problem. The uniqueness that I thought might be an advantage was also a problem as, unlike with repeat patterns, fabrics and wallpapers would have to be bought in lengths defined by the images. What beckoned, I felt sure, was a new 'Digital Age' of one-off, bespoke design – but that was difficult to pioneer from a converted garage outside Cardiff.

Alongside fabric printing I explored the potential of rugs and tiles. I knew about the former through a friend and distinguished mechanical engineer, Kerry Mashford, who had helped a well-known carpet manufacturer, Brintons of Kidderminster, refine a digital Axminster loom which they subsequently patented. The loom could weave-in any one of twenty-four colours

of wool at any point in the rug; the colours in turn were drawn from a standard library of over 600. It worked beautifully with several mineral/fossil 'designs' but presented the usual problem of 'getting to market'.

Thanks to another patented process ceramic tiles could be printed using transfer films made with a standard office laser printer – I say 'standard' but in point of fact it needed a little modification, of which the manufacturers did not approve. The printing powder, however, was far from standard: filling a cartridge for paper printing might cost £200; for ceramics it was £2,000. Happily, the small firm with which I was working wanted an eye-catching feature for a trade show. I suggested a piece of crazy-lace agate from Mexico, which forms flat in crevices rather than voids and typically has an appealing undulating profile. After being duly replicated in tile and exhibited, it was turned into a unique bench or table; with new 'dancing' legs it now sits in my living room.

In the spirit of the open-ended research with which I had been pursuing the 'data from nature' I decided to approach the local fashion school in Newport School of Art (then part of Newport University, now the University of South Wales). The tutors leapt at the idea and a second-year project – 'Frocks from Rocks' – was organised for the Spring term of 2007. The process was fairly traditional. The students developed a design using a cheap white fabric, which they called a 'toile', a term new to me. We explored the idea of projecting images onto this to adjust the scale of the scans, but most were happy to choose an image from a selection on Powerpoint. Several metres of each were printed, patterns matched to fabrics, and dresses sewn.

The results seemed to me appealing and I offered to sponsor fabric for one of the students, Sophie Shardelow, through her final year. She produced a striking collection based on the eminently 'mineral' theme of alchemy, using a beautiful piece of tiger's eye as her source. Some dresses from agate followed for a local exhibition – to me they somehow appeared ideal as off-duty wear for the female crew of the Starship Enterprise. After completing her studies Sophie

joined me as an assistant and played a decisive role in what happened next.

In November 2009 I was listening, as usual, to BBC Radio's flagship breakfast news programme, 'Today'. The principal presenter and famously searching interviewer, John Humphrys, introduced a certain Ed Burstell, the new Head Buyer for Liberty of London, then rumoured to be in dire financial straits. 'What are you, an *American*,' Humphrys asked with typical directness, 'going to do for this quintessentially *British* store.' Among Durstell's ideas was something that had proved successful in New York: an 'Open Call' to the 'great British public' to bring in their ideas and products. The first event was to take place a few weeks later and, at Sophie's insistence, I duly registered – albeit, after so many bruising rebuttals (I had already sent material to Liberty), not with great expectations.

As the big day of the Open Call approached I succumbed to a heavy cold, and – with cold, rain and strong winds forecast – decided to register for the next event, the following February. Two weeks before that I had a phone call from a television company: the day was to be filmed for a television series called 'Britain's Next Big Thing'. After being vetted by a buyer and journalist, interviewed by one of Britain's best known businessmen Theo Paphitis, a star of the famous 'Dragon's Den' for would-be entrepreneurs, I finally got to see Burstell and his assistant at 7.00pm. They loved the scarves and a first collection in Liberty beckoned.

Two weeks later I was being interviewed again by Theo Paphitis, this time in my home, which was by then turning into a veritable cabinet of curiosities of mineral delights (see chapter 5). A visit to the new printers – Maver in Como – followed and in July the first collection was launched. Burstell and his buyers liked what they saw sufficiently to order a further two hundred before a single scarf had sold. And sell they did. The scarf was the perfect vehicle for one-off digitally printed images and in time, thanks to the efforts of my business partner, Martin Price, 'Weston Scarves' found their way into other leading department stores in London and in several major cities abroad, as well as boutiques throughout the UK.

'Britain's Next Big Thing' was eventually shown in Spring 2011, in a prime-time slot on BBC2. It attracted a decent audience and led to the unfamiliar experience of being recognised by complete strangers. New outlets opened for the scarves and Liberty sold a range of other products – iPhone cases, silk-covered journals and even umbrellas – but other outlets for the manifold large-scale uses for the images were restricted to a commission from a friend of a friend in Edinburgh, prompted by a small exhibition in the Royal Incorporation of Architects in Scotland; a hanging for a narrow atrium; and two projects with the architects Patel Taylor, who were friends and already knew about what I was doing.

The Edinburgh project was modest but rewarding, a panel of printed silk to line the generous wall on the half-landing of the staircase of a house on the fringes of the city's magnificent Georgian 'New Town'. I suggested an image from moonstone which flashed with small areas of colour amid varied nocturnal blues and greys, and it was duly installed in three horizontal tiers. The client later wrote to say that both she and her husband were delighted with the result, which 'makes going to bed feel like ascending to heaven.'

The atrium was a narrow, tall space over the entrance to Cardiff University's new Health Library. The client selected one of my favourite images, a fluorite 'waterfall', which happily was large enough to print at the required four-story height.

The projects with Patel and Taylor were on a similar large scale. The first was to create a very narrow, three-storey road frontage for a new dwelling in London – the principal spaces lurked out of sight in the depths of the block and were adapted from one of their former offices. They chose an image from a beautiful level-bedded agate to make a three-storey-high screen of glass louvres. It was not then possible to print directly onto glass with ultraviolet-light-proof materials and so the image was applied to a delicate, translucent organza. The silk was laminated between two sheet of glass, the laminating films providing the requisite UV protection. When the louvres were being assembled passing cyclists and drivers were regularly stopping to see the novel sight. The house won the *Sunday Times* newspaper's 'Small House of the Year' award for 2012, but no further leads followed.

Later the same year the London Olympics were staged, and Patel Taylor were one of several practices invited to design apartments for the competitors' 'Village'. They used four mineral images on glazed screens to enliven the blank ends of the entrance foyers, but sadly only one – from an amethyst – was recorded, the photographer being told to leave by the security guards before he could complete his commission, such was the 'sensitivity' attending all things Olympic.

The scarves are still selling well in Liberty and elsewhere, and some interesting commissions emerged, notably to create designs for English Heritage based on the rocks of Stonehenge (below). We found some nineteenth-century thin sections in a Scottish museum collection, made from the monument itself, and even if the images were not the most beguiling, the historic source added lustre. Whereas I valued nature's infinite, subtle variety, the fashion world demanded constant and reasonably conspicuous novelty. The business was, in effect, living off a 'free' stock of data assembled over several years; to continually renew this stock direct from nature would prove both difficult and expensive. Part of the solution lay in using the mineral images for what they were, easily manipulated data-sets from which to derive new designs digitally. Given my 'devotion to nature' this didn't appeal: the business future of Weston Scarves lay with Martin, while I moved on to a different idea I called 'Digital Paints'.

4

From Digital Paints to Molly's World

By the time I presented the fourth collection of scarves to Liberty in 2011 I had amassed around 3,000 files of data from nature – and suspected that roughly 300 might pass muster as designs for scarves. Looking through the many that probably wouldn't, I was struck that a good few might work as means of colouring drawings. A few tests confirmed the potential and so I spoke to an artist-friend, Ruth McLees, who had extensive experience of working in Cardiff's schools, and asked if she would like to help. She recommended approaching Albany Primary School, a friendly, typical inner-city school housed in an Edwardian building with the kind of hard, tarmac playground I remember only too well from my own childhood.

I approached the head teacher, Angela Lepore, with the idea of running a workshop based on

drawing pet cats, with my own dear Molly as 'model' – in slides, not life. She warmed to the idea and decided to let both Year 4 classes of eight/nine-year-olds participate. The children had been well briefed – I was left in no doubt by our reception that the prospect of having both a professor (whatever that might be) and a tv-star (*sic*!) working with them was intriguing...

The resulting drawings were taken home and scanned, and around half received 'digital fur'. The results were presented to the children by being printed a little larger than life-size on a fifteen-metre-long 'scroll' – and seemed to me amazing. The quality of the drawings would have been lost if the children had coloured them in the normal way with crayons or paints. This is not to deny the value of traditional art-work, but the synthesis, as one of our External Examiners at the School of Architecture observed, of 'two of the purest things in the world, children and nature' was, and to me remains, remarkable.

Ruth and I returned to the school to run another session, which yielded dogs, dinosaurs and a few rabbits, and the idea of a 'travelling roadshow' for schools housed in a standard shipping container – a universal technology that

could be rented anywhere and relied on to be identically sized. On an unrelated project I had been experimenting with the then relatively new, high-powered neodymium magnets. An exhibition, I thought, could be rapidly fitted inside a steel container using them, and what emerged emulated the crowded intensity of a Renaissance cabinet. Particularly successful was a padded silk floor with 'rock' cushions, based on a specimen of Russian charoite. The intention was to provide a striking tactile as well as visual surprise.

A class of Albany children happily trekked across from their school and were delighted with the experience of drawing in this cramped enclosure. I left the doors open, fearful that they might find being locked in a steel box intimidating, but the first thing they asked was that they be closed, explaining that 'it's like a crystal cave in here'. Funding to tour the container was not forthcoming, but many of its contents are still available and I live in hope.

Another idea that began to germinate from the first experience of working with children was a digital app that would enable them to draw and colour digitally. At that time, to be honest, I barely knew what an 'app' was, let alone what

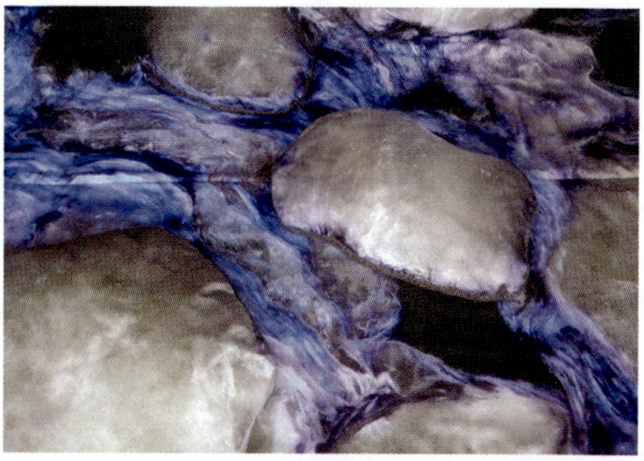

they could do: it was, after all, just over a year since the first iPad was launched. Little did I know that turning this hazy idea into a working app would come to occupy much of my time, of which more later.

With this long-term idea in view, I decided to make a further experiment. A former student, Rebecca Pike, happened to be doing some work at one of the UK's largest primary schools, Woodside Primary Academy in Walthamstow, and put to the head the idea of a lot of children contributing elements to a design for a scarf based on a wildflower meadow. This time we worked with six- and seven-year-olds, and instead of asking the children to draw from memory and imagination, which would have resulted in a meadow consisting almost entirely of daisies, we gave them printed 'models'. These were vector-traced from illustrations or photographs of two-dozen native British wildlflowers and were deliberately 'clinical' in character. The children were asked to make their own interpretations of these using pencil on paper. These were then scanned and we returned to work in small groups in the school's computer room, temporarily equipped with trial versions of 'Photoshop Elements', downloaded against several teachers' email accounts (the need for the app was ever more obvious!).

Again, the results were striking. Back in my studio the flowers, grasses and assorted 'meadow creatures' were composed into a two-metre-long design for a scarf (below), which was printed on a light, summery fabric consisting of 85% modal (made from beech-tree fibres) and 15% silk. The buyer at Fortnum and Mason with whom we worked on scarves thought it delightful and was all for stocking it. Sadly, I gather, she was overruled on the grounds that 'Weston' equalled mineral images, and this deviation would hinder brand recognition. In commercial terms this may have been right, but the experience only served to heighten my dislike of branding.

Alongside the work in schools I started to contribute to a privately run day-care-cum-art facility in my village, known as the Crafty Artbox. While working there I met a conspicuously bright young eight-year-old boy, Sam Boyce. With his and his mother's enthusiastic support I decided to test the limits of children's capabilities by lending him an old MacBook with Photoshop installed. I prepared some notes and gave him a half-hour

introduction to the software, and during the summer he completed, without any adult help (as his mother observed, neither she nor his father had a clue how he was doing it!), a beautiful meadow (below) and a collection of 'dragons' and planets that I put into a repeat design (pp.136-7).

Using the data files as 'digital paints' opened other possibilities. For Liberty, I made a small scarf collection based on silhouettes of trees (p.139). The trees were photographed using a pocket camera and the image then turned into a silhouette using the 'Threshold' mode in Photoshop. The same mineral image was then pasted into the black and white areas, one or other of which was 'inverted' to its complementary colours. The process worked

with a wide range of mineral images and, despite the digital processing, the results had all the vibrancy I valued in the un-manipulated files.

A variation of the process used to create the tree images was tested with classic Morris & Co. arts and crafts designs. It worked best with those, such as 'Cherwell', that had three colour-ways. Instead of combining two mineral images, I now used only one, pasting it into each colour-way and then rotating or flipping two layers. The design would generally be clear after these procedures, and the balance could be enhanced by small adjustments in tone or intensity.

Athough I have a small example (p.194, bottom left) of the *pietre dure* work discussed in Chapter 6, thought to have been made in Florence in the nineteenth century, it was only after considerable experience of working with 'digital paints' that it struck me how similar, in essence, the two procedures were. The data from nature could precisely match a typical panel from a cabinet of curiosities (opposite), and I find it strangely appealing that digital techniques can make universally available something close to a decorative tradition that was once accessible only to the very wealthy.

The need for an app was becoming urgent and after taking soundings locally I appointed Joe Offside of iApp Manager to work with me on an iPad/Android tablet app (to avoid clumsiness I will refer to both as 'pads'). We decided on three basic processes: drawing and colouring; montaging elements together to make a picture or design; and a block-repeat pattern-maker based on a square repeating 'cell'. Children, not surprisingly, love to draw directly on screen with their fingers but to achieve greater subtlety we made it possible to import drawings using the pad's camera. I called the app 'MollyPic' after its host, a drawing of Molly produced in the first workshop with Albany School.

The first examples of work made with the prototype app are encouraging – the 'magic' of the pattern-maker never fails to please and one incident confirmed my belief that, contrary to what you would expect from the flood of bright, often primary colours that seem to be the default setting in many cartoons and illustrations, children are admirably sensitive to subtleties of colour. A girl announced that she was going to draw 'one of those red mushrooms' and duly coloured it with a flat red. I asked why she hadn't used a digital paint and she asked 'is there a bright red?' I told her she would find some in the 'Flowers' folder. A dappled red was duly inserted and she sat back and said 'it looks alive now'. Indeed it did, just as the 'junk food' (his title!) that a boy produced at the rate of an item a minute looked edible. Molly's biggest fan soon emerged in the shape of a girl named Amy Churchlow: she designed her a delightful Easter bonnet and made her the star of the pattern for a dress.

Realising that children's attention-spans are not great and they might need further stimulation to make the most of MollyPic, I expanded the vision for the app to include an educational 'world' – to be called, needless to say, Molly's World. Her Highness is joined by her best friends and co-hosts, the dog Patch (to ensure gender and species balance) and the rabbit Thumper (to add a category-bending element of wildness).

The interface for Molly's World was inspired by an experience some twenty years earlier. My oldest and best friend from schooldays was then teaching German at Oakham school and I was living nearby in Leicester. He suggested I visit him to see an exhibition in the Goldmark Gallery in Uppingham of the work of Anthony Gross, the UK's leading etcher. As soon as I entered the gallery I spotted at a distance a 'must-have' and without close inspection asked the owner to package it lest it fall into other hands. Returning to Stephen's home his young daughters, aged

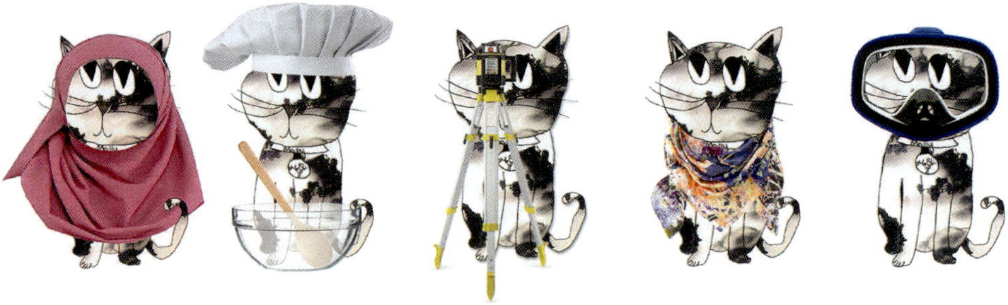

around six and eight as I recall, were eager to see my purchase. 'Look Sarah, here's one of those insects that can walk on water.' 'And look here Elizabeth, there are lots of small fishes and pebbles.' What to me had seemed reminiscent of an etched Jackson Pollock, all 'cosmic clouds' and other broad, richly textured formations, for them teemed with endless deftly caught details: it was, in point of fact, a vivid representation of a few square metres of a stream, albeit with the rather grand title 'Mississippi'. Children, I realised, have to learn to look at 'wholes', at 'compositions', and naturally immerse themselves in tiny 'facts'.

If the home page were to have similar qualities to the etching, it would be radically different to anything I had seen, so rather than plunge in with expensive coding I offered it as a final-year 'thesis' project for students of Computer Science at Cardiff University. There was one taker, Robert Kellow, a mature student with children of his own who did a fine job of making a version to run on a PC (and duly took a first class honours degree). Several friends were dubious about this radical departure from the norms of pages and drop-down menus – norms, I pointed out, with which children would hardly be familiar. Happily, Peter Trevitt, the former director of Cardiff's famous hands-on science centre Techniquest, was more encouraging. He asked if I knew of anything like it, which I didn't – and neither, to his surprise, did he, leading him to suggest that it could do for children what Apple's introduction of the virtual desktop had done for adults. I live in hope!

The interface for the launch version of Molly's World is illustrated below. It recognises that children are naturally inquisitive, fascinated by detail, intensely visual, and have small, eager fingers. The portals to MollyPic and other key elements, illustrated later, are ranged across the bottom and everywhere you touch leads to somewhere new. Patch's Balloon launches endless Journeys of Discovery in the Fields of Knowledge, the MollyJet which roars across when the app is opened is the invitation to a world of travel, and beyond the rainbow lie the microscopic worlds of Alciteland and the MicroScapes. The landscape, with its invitations to explore life on

It's from an agate. Is that a type of rock?
No Patch - agate is a mineral.
Exotic Cliffs
Richard Weston, 2015

Dr. Alison Banwell
Welcome to Antarctica! See its wild-life, hear ice crack, and enjoy Alison's reports.
Certainly is chilly down here!

m20 Bank of Molly
I'm sorry Mr Patch but there just aren't enough Mollies in your account to wallpaper your whole bedroom!
Molly Weston MANAGER

Come on Molly! It's lovely and warm!
And WET!

land and in the oceans, is fringed by Antarctica, presented – to the eerie sounds of cracking ice – by our very own correspondent, the glaciologist Dr Alice Banwell, formerly of the Scott Polar Research Institute at Cambridge and now of the University of Colorado.

Adult conventions – 'pages', drop-down menus, swipe-through galleries – have little place here: this is not a training for adult life (pity the children in countless primary schools mastering Microsoft Office!) but a world in which to become immersed, to learn naturally, and to create freely. I think of the 'pages' as 'places', not graphic layouts, and the ease with which data can be stored in the Cloud and handled by the app's Content Management System (CMS) offers diverse ways for children to engage with content and for us to personalise their experience. Much of the non-visual information will take the form of conversations between Molly (something of a know-all, needless to say) and eager-to-please Patch, presented using both printed and spoken words. This will make it more accessible to younger children and allow the intriguing possibility of having print in one language and sound in another, or vice versa – the app's potential not only to appear in other languages, but to help children to learn a second one is something we will be exploring.

In presentations I often describe Molly's World as 'an antidote to the National Curriculum', a remark which, following a local gathering of members of the Royal Society of Arts, led to teaming up with Steve Rees, the head teacher of Evenlode Primary School in Penarth, a close neighbour which quickly became the testing ground for the app. There is, I hope, a 'naturalness' about Molly's World and the responses of our focus group of eight-year-olds has been encouraging. They love the characters, who compared with most cartoons seem to me to have genuine 'character'. If children's art can have masterpieces, then Molly is surely among them: by turns mischievous, knowing and kind she seems to me to embody those feline qualities that endear them to their human devotees. We have so far made only one short animated story, in which Molly falls down a cliff and is rescued, thanks to Patch's timely efforts, by the MollyCopter. It is presented without speech, just music and written texts like the old silent movies. In the age of Pixar I was worried that the children

would find it too simplistic. How wrong I was. The fall happens early and abruptly and brought tears to the eyes of Molly's most avid fan who assures me she now watches it every day.

As for learning visually, this seems to have resonated more with early thinkers on children's education than those with the greatest ever means of marshalling visual data, the Web, at their disposal. As long ago as 1602 Tommaso Campanella envisaged a utopian *City of the Sun* whose plan echoed the new, heliocentric version of the Solar System. Residents were to be housed in concentric circular buildings with walls painted as a living encyclopedia of the arts and sciences: playing among them, he argued, children would learn naturally and without strain. Images were also central to *Orbis Pictus* (The World in Pictures) of 1658, a pioneering book for children by the Czech John Amos Comenius. Now recognised as the father of modern education, he argued that children learn best by 'obtaining ideas through objects rather than words.'

Children may love assembling words and images in Powerpoint and at that age it may not have the stultifying effect on thought patterns that has been argued about its use in organising lectures and business presentations. But personally I am not so sure: the medium and the message are often one, as Marshall McLuhan argued so long ago, and with digital media this has never been more true. What is certain is that there is more money to be made by writing addictive games than educational apps, so the dearth of enlightened options is not so surprising.

Convinced from my work with the digital paints that the children would be able to create things worth printing to make t-shirts, cushions, dress and shirt fabrics, even wallpapers, Molly's World includes a Workshop where children can order physical goods and a Bank in which to pay for them. This is in part an attempt to redress children's growing, according to press reports, lack of understanding of money in a world where most purchases are made digitally. The intention is that parents and grandparents will purchase credit for children in Mollies (the Molly tracks the £ Sterling and is worth 10p) and when purchases are in prospect the children will see graphically how they affect the balance of their account. In cases of serious overspend they will be asked to visit Molly in her capacity as Bank Manager (p.144) – given the details of the proposed purchase, the CMS will be able to personalise the message for each child. Some, I have no doubt, will happily spend time placing expensive orders simply to enjoy Molly's response!

Molly's World is a work in progress. It will be launched in July 2019 as 'Mollypic' and upgraded seasonally into the full educational app. Its adult counterpart, CreateForAll, will appear at the same time. For this we will be providing extensive in-app and online resources to enable adults to overcome that lingering, for many, obstacle to visually creativity: 'I can't draw'. The app will encourage them to try – it can turn surprisingly halting efforts into decent motifs for pattern-making – but also offer grey-scale templates that they can bring to life with the digital paints.

A smartphone app, ArtForAll, which forms part of MollyPic and CreateForAll, was launched (but not promoted) in October 2018. It enables users to blend a surprising range of digital paints – captured from minerals, textiles and the world around, or created with pastels – with a photograph taken on the phone. We offer regular downloads of paints, and users can create their own using the camera or the larger apps. Although primarily intended to be a relaxing and enjoyable way to make pictures, ArtForAll can transform banal subjects like road-markings (p.145), and rewards sustained efforts at honing one's eye. And as the examples show, it can yield images that would work well as cards or framed on the wall.

Molly's World began life as a simple colouring app but has grown into an educational world and spun out creative tools that enable adults and children alike to participate in the Digital Revolution as creative producers not just passive consumers. It is being developed through a non-profit Community Interest Company and we hope to enlist users as subscribing 'Friends of Molly': for the price of a cup of coffee per month they will have access to the apps, tutorials and sources of inspiration.

Art
For
All
ArtForAll
Save

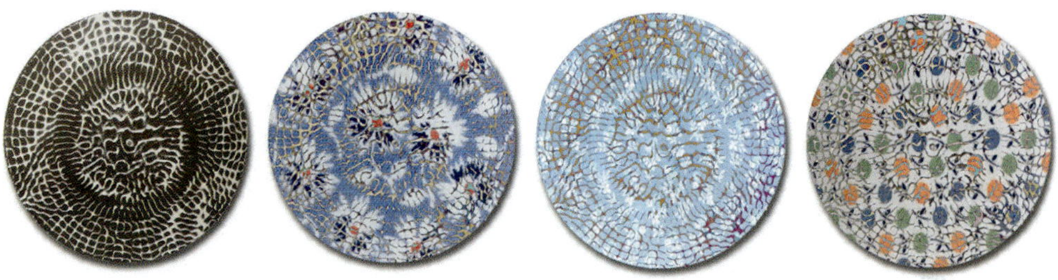

Mollypic's lanch project is a digital reinvention of a traditional summer pleasure: beachcombing. Cast as 'good pirates', Molly, Patch and Thumper set sail from the Chalcedony Coast to roam the Pirate Seas, from where they bring back treasures and bury them on beaches, to be discovered by children using smartphones. Thanks to their wealth, the pirates have scattered their islands with replicas of monuments from around the world: some can be seen from the ship, while others are found by launching the MollyDrone. Presented in an interactive video they can be touched to reveal more.

Everywhere on the landing page (opposite) is active and leads to features to be introduced over the course of the summer. The three clouds lead to postcards from Molly and Patch (who will be holidaying in the USA in August); short reports from video correspondents around the world; and a gallery of work made using the app. The distant cliffs reveal the animated story, 'A Cliff Too Far'; Patch's balloon leads to his 'Journeys of Discovery'; and finally, the floating mollypic letters (animated to move slowly with the waves) take you straight to the app's creative heart. Children remember these locations easily and love searching for surprises if they forget: titles are not required. Some content is illustrated below, and will remain available as new elements are added from season to season.

Register here
mollypic

CreateForAll offers similar creative tools to mollypic, with a few more advanced functions suited to adults, numerous tonal templates for colouring and access to a 'Friends of Molly Room' with its blog, tutorials and sources of design inspiration.

Through its montage and pattern modes it is designed to enable users with only modest drawing abilities to create high-quality everyday goods such as greeting cards, ceramic tiles, printed t-shirts and fashion and interior fabrics.

5

LABORA-TORY: DOMESTIC EXPERI-MENTS

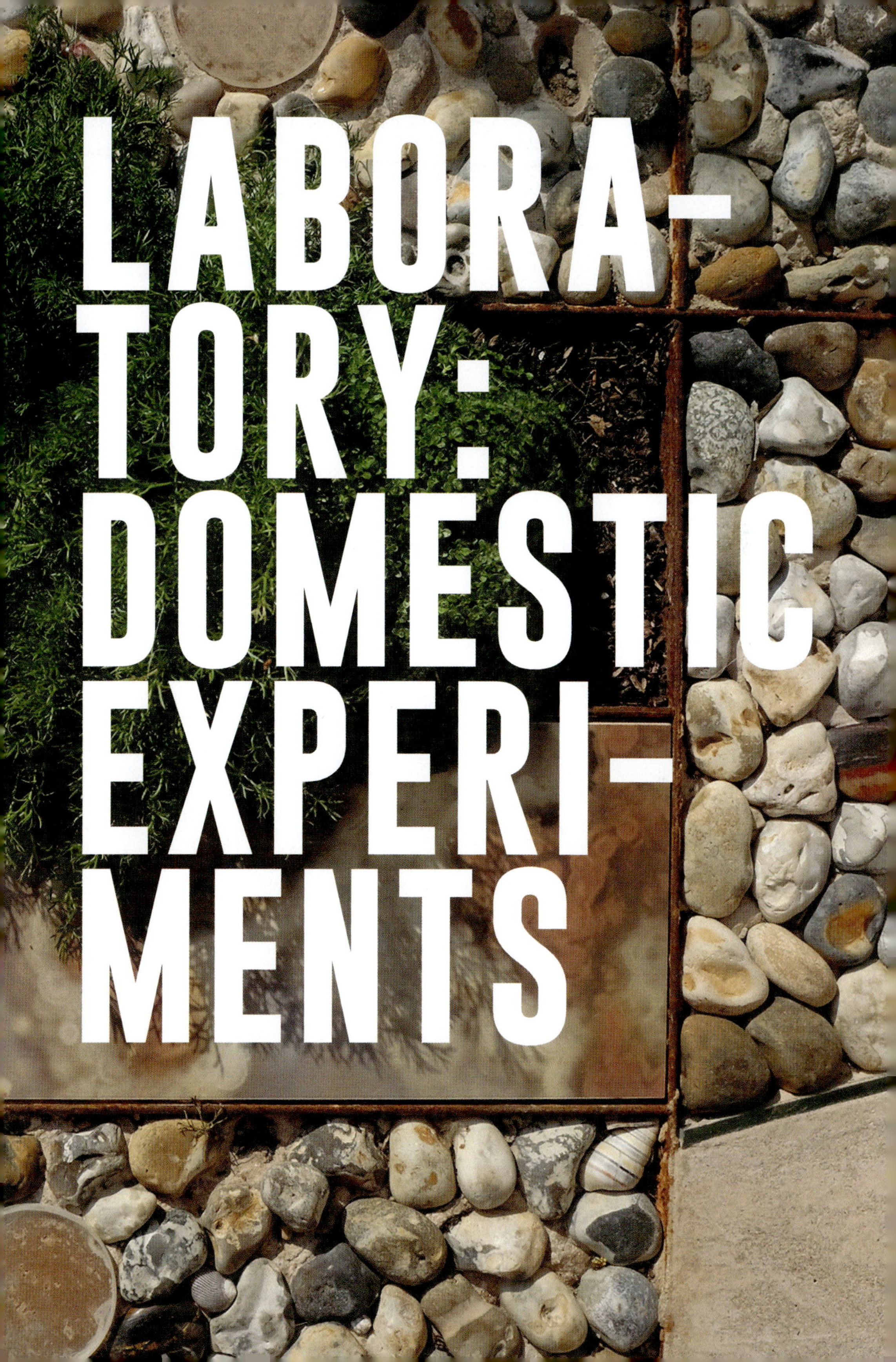

With little outside interest in the design potential of the data from nature I decided to use my house as a 'laboratory'. Courtesy of the ever-growing mineral collection it was turning into an informal cabinet of curiosities – in addition to slices for the scanners I couldn't help acquiring conventional specimens – and it soon it became a test-bed for digitally manufactured materials. When I began, these were limited to fabrics and tiles, and the latter were insufficiently resilient on either floors or externally. I still had access to the digital loom at Brintons, so a rug was an early addition. I chose an image from Australian variscite that reminded me of an aerial view of the Earth's surface – and in the nick of time. Shortly after, the company declared the loom off-limits to outside designers.

The transformation of the ground floor began with the installation of a high-level book/display shelf, with short runs of full-height shelves – 'strata' as I thought of them – appearing at intervals. To line the blank living/dining room wall, and create a shallow zone of storage, I introduced linen-covered panels printed with two images, one from paesina stone, the other from quartz. The former was chosen partly to 'rhyme' with the Virginia creeper growing on a two-metre length of brick wall that screened the neighbour's drive. After a while I decided to extend the quartz image into the rear garden as a tile panel. Tiles were not then guaranteed frost-proof but the manufacturer thought I might get away with it in the mild climate of South Wales. Visually it was not a great success, the joints in the tiles proving more intrusive than expected, destroying the hoped-for 'continuity'. And technically it was a disaster: my corner of south Wales duly had its coldest winter for decades and the surfaces of several tiles were badly spalled.

Living room curtains were printed with the same image from moonstone used in the house in Edinburgh. And to greet Theo Paphitis and the *Britain's Next Big Thing* film crew the dining room sliding doors received a large roller blind from mica – in inverted colours it suggested, to me at least, a kind of mineralised snow storm. Briefly recorded in words, all this may sound like a recipe for visual dissonance, if not chaos, but to my eyes the varied colours and textures harmonise admirably. During the 'Frocks from Rocks' project several students commented to me that their mineral prints contained colour combinations that they were taught 'don't work' – but did. It is, I think, a universal 'fact' of nature, for

which I have never heard an explanation, that every 'colour scheme' no matter how vivid and varied – as, for example, in a coral reef – seems to form an harmonious whole.

I had bought a large-format digital printer soon after beginning 'the scans' and later decided to add a laser cutter to explore other uses for the data. I was keen to see how its ability to track intricate contours might work with vector-tracings made direct from natural sources. My favourites were snowflakes made from clear acrylic sheets cut with files traced from the pioneering photographs of Wilson Bentley. The intricate cutting involved was uneconomically complex: it might take up to twenty minutes for a crystal five centimetres across, but the results captured the 'natural variation' in a way that made a mockery of their

distant, injection-moulded cousins used as Christmas decorations. Crucial to the visual quality was the way the cut edge – consisting of multiple facets rather than a flowing 'line of beauty' – caught the light. I decided to test their 'naturalness' by applying some to my living-room window, and was convinced they passed muster when a friend asked, after being sent a photograph I called 'Window Weather', how I had managed to photograph snowflakes in the rain!

The laser cutter was also put to work etching grass, fossils – which had a few friends, uncertain about the geological origins of slate, thinking they might be 'real' – flowers and moths onto slate and uncoated cardboard. The most convenient source of slate was coasters and placemats stocked in a well-known supermarket, and 'Place Mat' acquired a new meaning when I etched onto

one (sadly now lost) an historic Ordnance Survey map of the village where I live. I thought there might be a modest market for similarly bespoke designs until discovering the cost of buying the 'rights' to use the OS data commercially.

The most extraordinary result with slate came in error. Instead of the usual silver-grey colour of the etched lines, a network of raised, golden ones appeared. I quickly realised that the laser had been set on cutting strength by mistake, but was still baffled by what could be happening. I tried it on other slates but it was only the Chinese material that responded so dramatically, Welsh slate in particular being almost completely resistant to the heat.

Armed with this information I took the sample to show my geologist-friend, Alan Channing. After a few moments he said he thought he knew what was happening. Typical Chinese slate, it transpires, has about a 70% water content (not obvious to the layman!), which he suggested was being boiled near the surface by the laser, destroying the aeons-old structure of the stone and drawing to the surface our old friend pyrite, which was then re-crystallising as a raised, 'bobbly' line. Seen in sunlight the effect was exquisite and ought to be pursued on a larger scale: the pyrite is fragile and couldn't be walked on, or even used for table settings, but panels would work beautifully on well-lit walls.

Cardboard, not surprisingly, did not offer the surprises of slate. But, once the optimum intensity for the laser was found, it could record exceptionally fine detail and had a character all its own. With all colour drained away the images feel strangely melancholic, like the ghosts of pressed flowers in a Victorian album.

An enquiry about making some unusual and luxurious candle-holders led into the world of an altogether more powerful technology – water-jet cutters capable of slicing, so I was assured, through 100mm or more of solid steel. I selected a group of conjoined circles in a piece of orbicular jasper, from which a double and single cylinder

could also be cut from within the larger cluster. The results, while visually appealing, proved too expensive even for a prime London retail location. They worked well enough as candles, but were at their most beautiful before filling, the play of light seeming, in places, to transform the curving glass walls into light-filled voids.

Following the purchase of the microscope, more of the lump sum from the pension was put to work. First came a plywood 'Staircase of Curiosities' to store the mineral collection. The design was based loosely on traditional space-saving stairs in Japanese Minka farmhouses, but in place of cupboards facing into the hall/room I opted for drawers in the risers. The design was unashamedly 'geological', spreading out at its base and returning along the hall with display boxes of varying length and height: I knew my cats would love the 'perching possibilities' offered by its various strata, and they have not disappointed. I found a superb craftsman, Wayne Tutssel, to make it: originally apprenticed as a cabinet-maker, he deployed skills and an attention to detail that are no longer easily found.

Next, it was out with the old furniture and in with a collection of Eames pieces that still seem to me, over sixty years on, as good as modern furniture gets. The sofa was retained but re-covered with a print from a favourite level-bedded agate. Leaky double-glazing, front and back, was replaced with new sliding doors, releasing a 'flow' of space – as we architects like to call it – that necessitated replacing the solid, zigzag support of the crazy lace agate-tiled table with

a set of slender 'dancing' legs. It was a nod to architectural fashion that I generally eschew, but at least here the leaning rods (created by rotating identical pieces of stainless steel with chamfered ends) do serve to prevent the top from racking.

Several years after my first efforts with digitally printed tiles, new 'potions' from the same German chemist made their use on floors and even outdoors perfectly feasible. A new kitchen was part of the 'works' and its floor was duly covered with porcelain tiles printed with an image from a 180 million-year-old sea-floor rich in ammonoids (the palaeontologists' catchall term for ammonite-like creatures) – an appropriate counterpoint to the 'sea and sky' kitchen roller-blind that, I like to tell visitors, is the only sea-view in my inland village. Glimpsed from the entrance at the far end of the hall it is almost believable.

Finally, I decided to tackle the front garden, which faces south-west and is far sunnier than that to the rear. The house, one of eight in a single-banked close with a solid screen of evergreen trees opposite, is set nearly a metre below the road. Privacy is not an issue and it made every sense – as several neighbours have done – to turn it into somewhere to sit out. And what better to sit out *by* – or look out onto – than a pond? I had wanted one since childhood and finally the opportunity beckoned.

To act as a low retaining wall for the neighbour's drive I decided on a 'Steel Cliff' made of three layers of zigzag-folded steel sheets slotted together. I had made similar 'slotted' structures from thin cardboard in one of the first projects of my architectural education: it seemed the 'natural' way to make something strong from thin, flat sheets and although they were eventually tack welded they would have been stable without.

The layers of the Steel Cliff diminish in depth as they rise, and turn inwards at right angles to support shallow terraces of planting across the garden, facing back to the house: the lowest rises straight from the water, while the upper is slightly higher than the now-invisible road. The planting was designed and installed by a friend and Royal Horticultural Society Gold Medal-winner, Vicki Wade. It is based on the kind of early-successional plants found on post-industrial sites, and a PR account might suggest that this is a nod to Wales's only remaining steel works a few miles away at Port Talbot. It crossed my mind but wasn't the real motivation: I wanted to celebrate the beauty of rust, on which John Ruskin had written extensively, and find it a perfect counterpoint to the greens of plants.

The visual pleasures of the planting come mostly from the tones and textures of foliage, with isolated, short-lived bursts of colour. The flat steel planes might almost be neat vertical cuts through clay-rich soil not – as anxious neighbours feared – alien industrial intrusions into a domestic environment. The steel is ordinary mild steel, rusting naturally. The cost is a third of the Corten used on buildings and the fabricator, Peter Gibbons of Printmet, offered the helpful advice that it should outlast me! The resulting colours also seem to me to be more subtly varied than is typical with Corten.

The data from nature, needless to say, had to be incorporated and I decided on a series of 'stepping tiles' printed with microscope-images from a Moroccan agate: several evoked rushes and other marginal plants emerging through water into view, an impression created by inclusions of linear crystals growing through the agate. To add animation and help with aerating the water I commissioned a 'Glass Volcano' from Rodney Bender of Innovative Glass Products. It was made by layering small shards of glass – recycled waste processed through his crusher – onto a shallow cone of sand in a kiln. This was then heated to 800°C, fusing the glass into an intricately varied surface that reflects light beautifully and looks even better with water 'erupting' from a central hole and flowing down erratically across its surface. The water then drops a few centimetres into the pond, creating sufficient ambient sound to complete the feeling of a private world and – an unexpected delight – producing a complex play of rippling reflected light across the angled planes of the 'Cliff': it can be mesmerising to watch and I must put a short video on my website.

For the patio – the Spanish word seems appropriate given that it is where the tiles were digitally glazed – I decided on a 'layering' of variously scaled references to nature, ordered by a shallow rectilinear grid of steel compartments. The presiding metaphor was a beach, and as on

a beach, water-washed pebbles are graded from small by the 'shore' to larger by the 'land'/house. The pebbles are scattered with occasional shells and small jaspers – perfect for children to seek out – and the pond-edge is lined with an image from a Hungarian agate that evokes an aerial view of a coastline. The full agate image 'lurks' beneath the entire patio, surfacing occasionally as isolated tiles, which mingle with 'slices' of planting and a network of concrete rhomboids awaiting the possibility of in-situ printing with images from calcite. Finally, the whole is scattered with small, apparently randomly placed, disks of glass: in fact, slightly self-indulgently, they follow the positions of the high magnitude stars above the house at midnight on my birthday (such star maps are easily plotted from online resources and are one of the most accessible sources of truly random arrays). In low winter sun, the glass circles create a slowly shifting constellation of stars on the living room ceiling, complemented at times by shimmering light reflected from the rippling surface of the water.

I was worried the patio might feel overworked, but having lived with the garden for nearly four years I find this intricate foreground works well with the broad planes of steel and subtle interplay of different greens and leaf forms in the planting. As with all gardens the year is punctuated with seasonal pleasures: the flowering of the yellow flag irises, arrival of mayflies and later dragonflies, the succession of waterlily flowers rising magically 'from the deep'. And it occasionally offers completely unexpected delights, such

as the breeding pair of ducks who have dropped in on several occasions. Despite worries about predatory herons, the goldfish have thrived and multiplied – the cats love to drink the water but have shown only a contemplative interest in these strange water-dwellers.

Small though it is, for me the garden has breathed new life into that old ideal, most vividly observed in Islamic traditions, of the garden as an earthly paradise shaped around life-sustaining water. As a synthesis of animate life and vital matter, 'analogue' materials and digital prints, it also hints at the 'manifold and multi-faceted' vision of architecture that motivates my ongoing research with data from nature.

6

HISTO-RICAL ECHOES

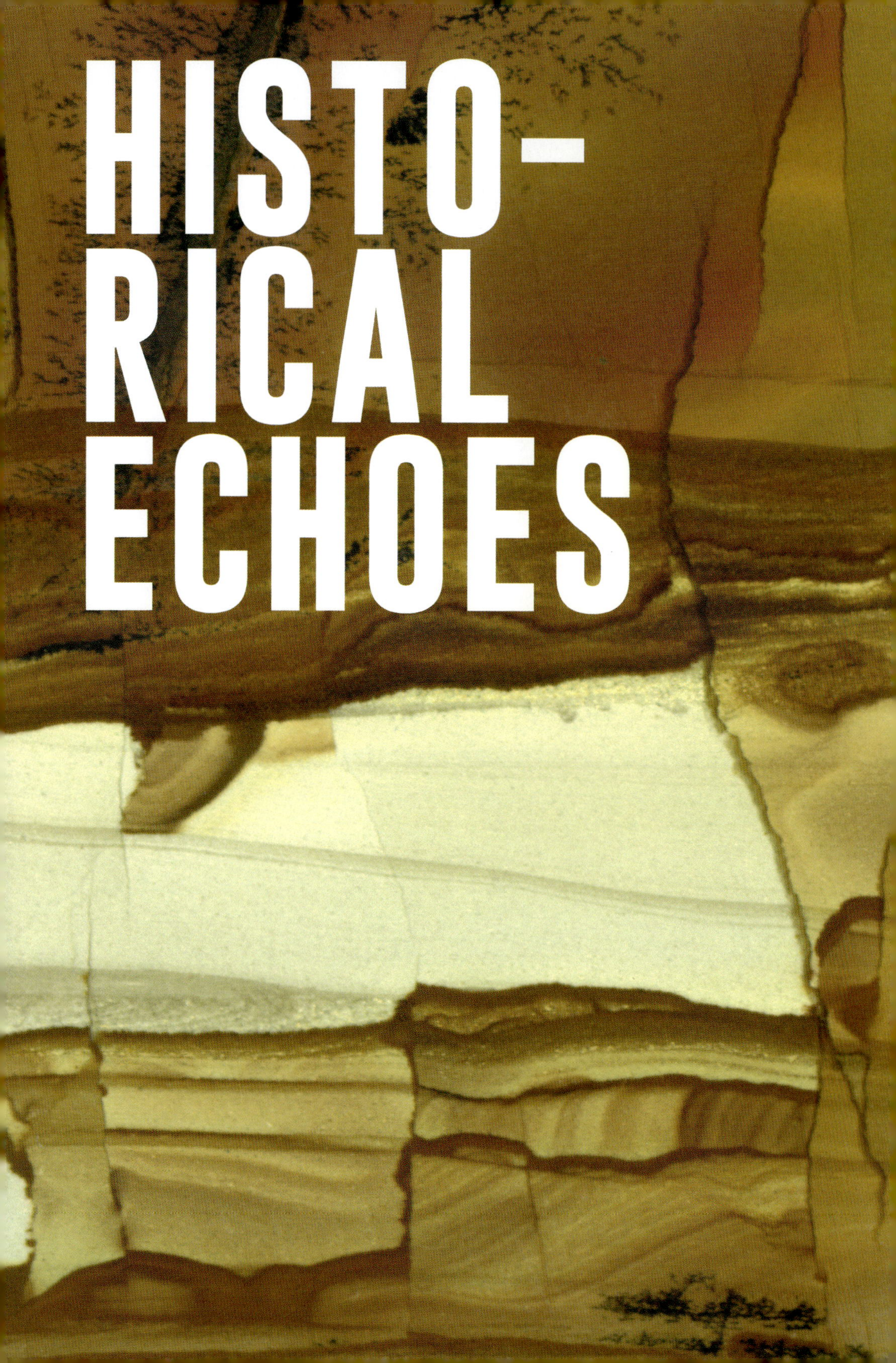

Among the thousands of scans I have amassed, a detail (p.180) from a large specimen of orbicular jasper holds a special place. It wasn't one of those that seemed obviously pictorial, but an image immediately came to mind: it seemed like looking up at the ceiling of a Baroque church, with the intersecting spheres suggesting architectural mouldings or stylised clouds framing a glimpse of the heavens – not the biblical Heaven of the Assumption of the Virgin or Christ in Judgement, but a glimpse of the distant universe captured by the Hubble Space Telescope.

I had a small ceramic tile made from the file and look at it most days. I still find it beautiful and slightly mysterious, but its significance for my unsystematic research came a few weeks later when I came across something strikingly analogous on the Web, a painting (opposite) of The Last Judgment by a minor German master, Johann König, known mainly – if Wikipedia is to be believed – for his painted copper panels. What made it special is that it was painted on agate, and is part of a magnificent cabinet of curiosities presented to Gustavus Adolphus of Sweden, of which its German maker, Philipp Hainhofer (1578-1647), modestly said 'some consider it to be the eighth wonder of the world.' It is now on display in the University of Uppsala.

I didn't particularly like the rather pretty rendering of an awesome event, but was fascinated that my jasper could suggest a similar image across the centuries to that conjured up all too literally by König. With helpful hints of blue sky painted into its cusped interstices, the agate's banded formations were readily subsumed into their figural role as clouds. At this point in my journey into the world of minerals I had not encountered paintings on minerals or stones and knew next to nothing about the world of cabinets of curiosities, which were at the time coming into vogue in contemporary art. The gap clearly had to be filled.

Cabinets of curiosities, it turned out, began life in the early sixteenth century as rooms, later turning into expensive cabinets like this seven-

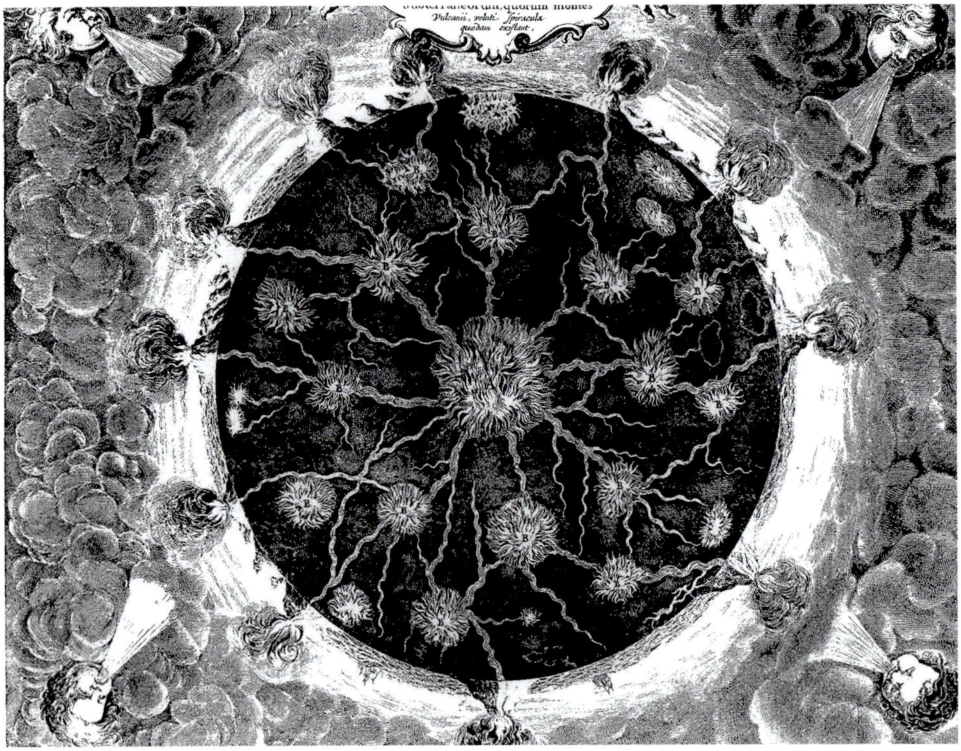

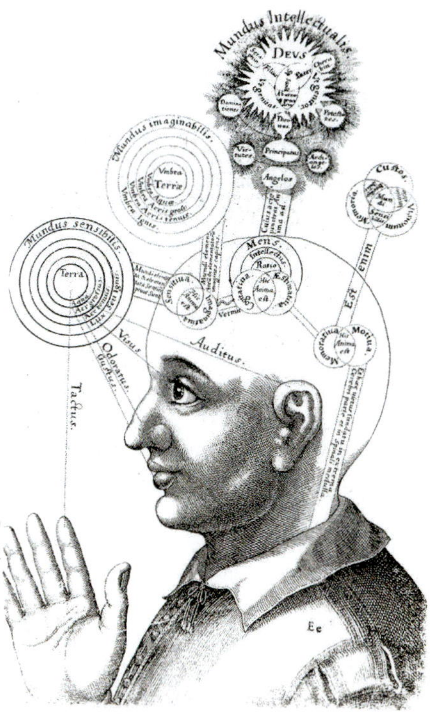
Mundus Intellectualis
Devs
Mundus imaginabilis
Umbra Terræ
Mundus sensibilis
Terra
Mens
Intellectus
Ratio
Visus
Auditus
Odoratus
Gustus
Tactus
Est enim
Memorativa
Mortua
Vermis
Angeli
Principatus
Ee

MUSEI
WORMIANI
HISTORIA
LUGD· BATAVORUM
EX OFFICINA ELSEVIRIANA
ANIMALIUM PARTES
CONCHILIATA
VARIA
TURBINATA

teenth-century example from the workshop of Melchior Baumgartner (p. 176). In the room-size cabinets were displayed private collections of items from natural history, geology, archaeology, religious or historical relics, small works of art, and Classical and other antiquities. Alleged unicorns' horns cold be found alongside fantastically-shaped corals, exquisite minerals next to stuffed animals (crocodiles seem to have been particularly prized), natural freaks like small creatures born as conjoined twins jostled with pieces of intricately carved ivory, fossils with insects preserved in amber.

Although they have come to be seen as precursors of the modern museum, the aim of the cabinets seems, according to the German art historian Horst Bredekamp, to have had less to do with the systematic ordering of objects into discreet 'departments' than with promoting comparisons between apparently disparate things. In some cases, even the division between natural (*naturalia*) and man-made (*artificialia*) was considered far from clear – life-like automata were highly desirable features of later cabinets – and as Enlightenment science began its inexorable march into specialised disciplines (physics, chemistry, biology, etc.) the cabinets reflected

the lingering Renaissance belief that it might still be possible for an individual to know everything.

Few things in the cabinets appear to have provoked more speculation than images to be found in what were known as 'pictorial stones'. And what images they were! A flint was said to bear a picture of the Virgin, and a stone found near Tivoli contained the monogram INRI (Latin for the inscription 'Jesus of Nazareth, King of the Jews', placed over the crucified Christ by Pontius Pilate). Italy supplied stones to the whole of Europe and several of its museums were rumoured to include exhibits showing a cross in feldspar, a majestic bearded head in alabaster, and depictions of the burning of Troy (marble veins were routinely interpreted as flames). In Flanders, allusions to Classical antiquity were replaced by biblical cataclysms, but the visual configurations in the minerals that gave rise to vivid but diverse interpretations were much the same.

The Jesuit scholar and polymath, Athanasius Kircher, provided the most comprehensive inventory of such wonders in his 1664 book, *Mundus subterraneus*. Inspired by a journey into the crater of Mount Vesuvius, this is best known for its highly imaginative illustrations of what we might call 'dynamic Earth' (p.178). Alongside his sometimes precocious geological insights Kircher catalogued stones using, as it seems to us, a system closer to a Borgesian fantasy than serious taxonomy. His categories comprised geometrical figures and letters of the alphabet; celestial visions; the terrestrial world; animate creatures; and religious figures. He also specifies chemical formulae to enable applied drawings to penetrate the minerals, leading to the suspicion that many of the so-called 'pictorial stones' may have been faked or at least 'chemically enhanced'.

Faked or not, such imagery was of long-standing. In the Christian world, for example, there is an ancient tradition of finding natural images of the faces of Christ or the Virgin Mary in book-matched marble slabs, and most of the 'genres' Kircher presented had already been described by Greco-Roman lapidarists, whose writings established a tradition that was elaborated in the Middle Ages. Agate, in particular, was appreciated for its 'varied shapes/either

of trees or crowned kings/or well-dressed damsels/and the forms of savage beasts.' In the mid-seventeenth-century Cardinal Richelieu's librarian took up the same themes, emphasising minerals' fantastic qualities, and as the scientific understanding of the Earth and its rocks and minerals grew, the world of the lapidaries became curiouser and curiouser. A 1778 catalogue of the newly established British Museum lists 'A Rough Egyptian Pebble' – in fact an agate – 'on which is a striking likeness of *Chaucer*, the father of the English poets.' At Strawberry Hill visitors could ponder, in another agate, a compelling likeness of Voltaire.

Among collectors of the late Renaisance and Baroque periods, no type of stone exerted greater fascination than the so-called 'Florentine marbles' which we now know as paesina stone and have already encountered. Found in hills north of Florence, these are cut from a fine-grained limestone which takes a beautiful polish. This brings out the varied colours and markings created by the ingress of water-borne elements between the bedding planes and through a filigree of exceedingly fine cracks produced – as we noted in Chapter 1 – by the collision of the African and European tectonic plates.

In his notebooks Leonardo da Vinci famously advised painters in search of inspiration to 'look at a wall spotted with stains, or with a mixture of stones, if you have to devise some scene, you may discover a resemblance to various landscapes, beautified with mountains, rivers, rocks, trees, plains, wide valleys and hills in varied arrange-

ment; or again you may see battles and figures in action; or strange faces and costumes, and an endless variety of objects, which you could reduce to complete and well-drawn forms. And these appear on such walls confusedly, like the sound of bells in whose jangle you may find any name or word you choose to imagine.'

Most paesina stones require rather less imagination to read pictorially, typically presenting what appear to be ruined settlements or weathered cliffs, often overgrown by tiny manganese dendrites. To reinforce the 'landscape' quality, wavy, twisted lines frequently pass muster as clouds, or a blue-green colouration suggests marshy ground. The fineness of 'nature's work' is startling and a nineteenth-century French history of minerals adopts the tone of an art historian in describing 'these charming accidents, in which nature so well imitates the productions of art,' noting that larger specimens frequently feature 'what painters call a terrace, a strip of ground strewn with shrubs and undergrowth which completes the resemblance to a painted landscape.'

Not to be outdone by Italy, England eventually acquired its own landscape stone with the discovery of Cotham marble: named after a village near Bristol, the geological formation involved runs through into south Wales. Unlike paesina stone, its characteristic 'hedgerow' formations are organic rather than inorganic in origin, recent research confirming the involvement of algae.

A substantial international trade in paesina stones flourished in the sixteenth and seventeenth centuries. Like many of its kind, the opulent cabinet presented to Gustavus Adolphus with which we began this chapter was partly decorated with them. They also entered public museum collections: mounted on slate and elaborately framed, they were presented to the public as natural works of art. Given their pictorial qualities artists began to use them as the 'ground' for miniature paintings in a similar spirit to König's *Last Judgement*. But whereas agates, marbles and jaspers typically offered only generic hints of large-scale natural formations such as billowing waves or clouds, 'the paesine,' as Roger Caillois has written, 'with their suggestions of cities or rocks are quite another matter. Here the painter may supplement the buildings with little black rectangles representing doors and windows. He may also introduce trees, though he need not put in grass or undergrowth: the fine dendrites of manganese provide an excellent illusion of these.'

Most of these paintings on paesina stone may seem to the modern viewer to amount to little more than pleasant divertissements, but in some the close alliance between stone and figuration raises the game to another level. A superb example, in the museum of the *Opificio delle pietre dure* in Florence, was inspired by Dante's Divine Comedy and shows Dante and the Roman poet Virgil visiting Hell (p.182). Against the dark ochre background, the naked bodies of the damned are indicated by cursory pale marks suggesting, as Caillois observes, 'a clear case of complicity between the subterranean levels of suffering and a stone that itself comes from the depths of the earth.'

Such complicity, the art historian Christopher Nygren has recently argued, turns 'on a peculiar confederation between the material substrate of the picture and the lithic resonances embedded in the stories they portray. ... Stone afforded the artist something unique, and the petrified material participated in the visual economy of the image, helping it attain a heightened sense of analogy between subject and object. These images present rather than represent.'

In a detailed analysis of a painting on lapis lazuli of *Perseus Liberating Andromeda* (p.182), made around 1595 by the baroque artist Cavaliere d'Arpino, Nygren describes how, in the given pictorial context, the lapis is transformed into an image of the sea, 'gently covered with floating algae and cresting waves', while a crack running through the mineral is incorporated to inject greater drama into the story. 'Andromeda's toes nestle up against this crack,' he points out, 'and her foot and knee seem to instinctively pull away. This motif suggests that the stone outcropping upon which the maiden has been placed is collapsing under her feet. She fears not only the

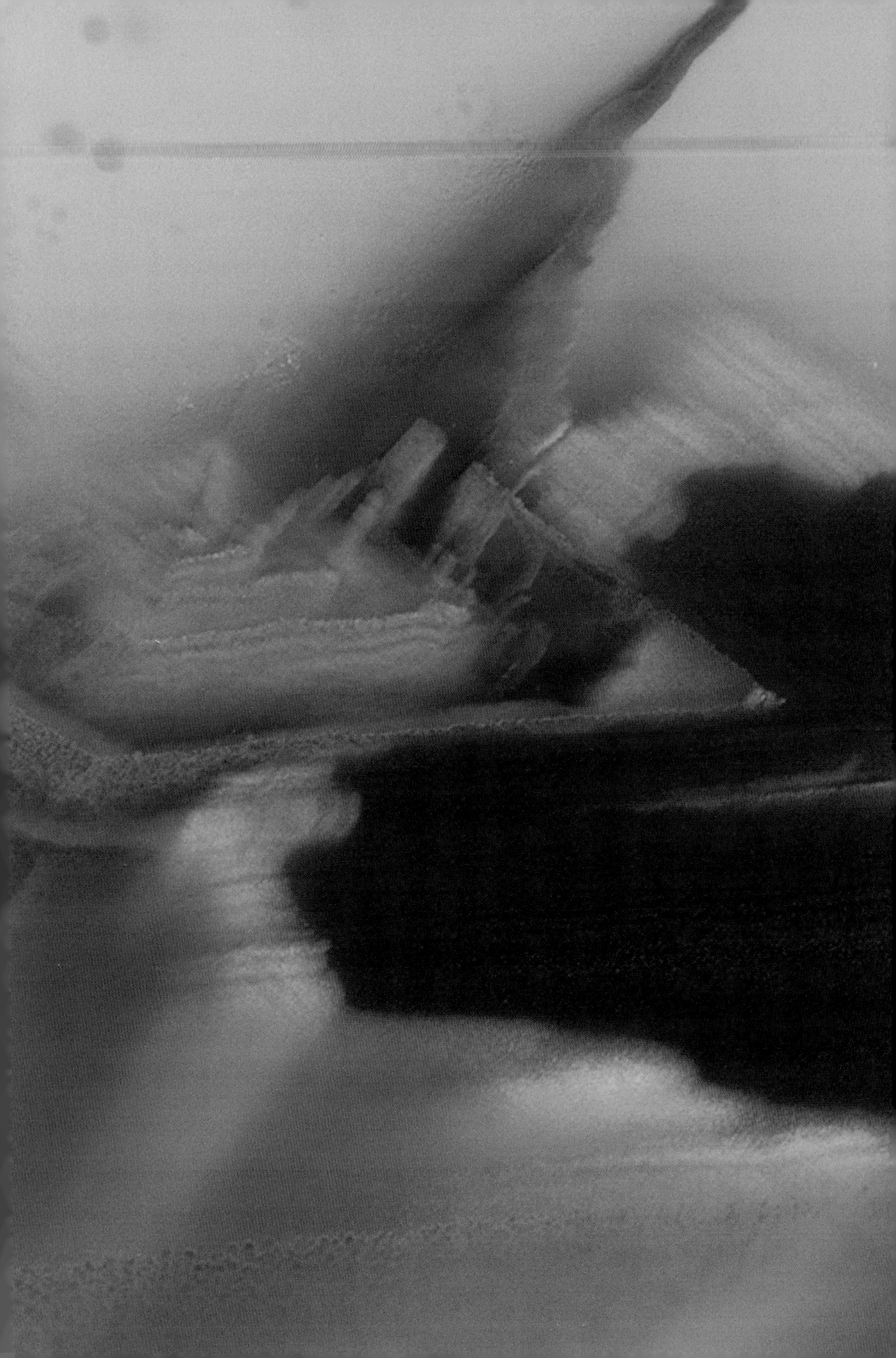

sea monster, but also plummeting off the rock precipice into the blue-green sea below.'

Unlike paintings on a canvas or a wooden panel, in which the support plays no active role in the work, with paintings on stone the relationship between figure and ground is unclear: where do the colours of the stone end and those of the artist begin? For the sophisticated owners of cabinets of curiosities, and those privileged to view them, such works tantalisingly challenged the distinction between *naturalia* and *artificialia* – much as, Nygren argues, paintings on stone 'challenge the hard ontological distinction between intentional beings and inert matter, which has been central to post-Enlightenment Western epistemology ... only by recuperating a pre-modern understanding of the vibrancy of matter can scholars fully appreciate the presentational.'

We will return to these ideas in the Conclusion. Now, I wish to turn to another aspect of the engagement with stones and minerals we

find exemplified in – or rather *on* – the cabinets of curiosities, which reflects a longstanding interest in their decorative possibilities. One of the strangest texts to come down to us from the Renaissance bears the imposing title *Hypnerotomachia Poliphili* (known in English as 'The Strife of Love in a Dream'). Generally attributed to a Dominican priest named Francesco Colonna, it relates the efforts of Poliphilo to win the love of Polia. As his name implies, the protagonist's love was not easily contained, and he proves equally passionate about architecture and antiquity. In his dream he constructs and visits imaginary palaces, temples and theatres: the woodcuts illustrating them are complemented by long, ecstatic descriptions of their material delights.

In the residence of Queen Eleuterylida, for example, Poliphilo passes through a court with a floor paved 'like a chess-board, alternately of coral-coloured jasper and of bright green with sanguine spots (presumably a form of heliotrope or bloodstone).' Surrounding this was 'a noble interlaced design of jaspers, prases (a leek-green quartz), chalcedonies, agates, and other conspicuous kinds of precious stone.' The seats are of 'red and yellow sandalwood' upholstered with 'green velvet', and the surrounding walls were 'all covered with plates of pure, lustrous gold'. And so the descriptions go on, for four pages in all, taking in lapis lazuli, vermilion, amethysts, garnets, emeralds...

Written during the Renaissance and illustrated with loosely Classical buildings, these descriptions have two roots. Firstly, the tradition of mosaic decorations, which the Romans derived from the Greeks and developed using both traditional tesserae and larger, shaped pieces of stone: known as *opus sectile* this was particularly widespread in late antiquity. And secondly,

the Byzantine/medieval love of surface elaboration, often using small tesserae of coloured glass. The Renaissance passion for *ornato* – meaning beautiful, precious or magnificent, and typically expressed through surface polish, embellishment and refinement – had much to with the preceding idea of *ornatus*, which to the medieval mind connoted God's transformation of formless chaos into a living world. Richly coloured stones (not those containing mysterious 'figures') were seen as representatives of the divine majesty. Writing in the sixth century, the mysterious figure of Pseudo-Dionysius the Areopagite likened coloured marbles in a pavement to flowers in a springtime meadow, but whereas the natural growths wilt and die away, he observed, the chapel 'shines eternally with its immortal flowers.'

It is understandable that we emphasise the discontinuities between the medieval and Renaissance periods, but it is well to remember that medieval religious thought was also deeply touched by Classical philosophy, notably that of Aristotle, and that craft traditions could hardly be transformed overnight. To take one example: worshippers in many of Rome's churches walked on pavements made in the richly ornamented style developed by the Cosmati brothers that flourished over several generations as a family business in the late Middle Ages.

In a comprehensive re-assessment of Renaissance aesthetics, Hellmut Wohl notes that 'as we read through Renaissance writings on art, it is not uncommon to find greater responsiveness to the aesthetic properties of materials, especially of coloured marble and mosaics, than to works of painting and sculpture.' The appeal of fine stones was almost universal, and in early Renaissance wall paintings, beginning with Giotto's cycle in the Arena Chapel in Padua (p.183), the representation of coloured stones and minerals takes up a surprisingly large amount of the available wall space. Even Leon Battista Alberti, leading advocate of the purity of white in churches, still commended the judicious use of inlaid marble and glass. For most Renaissance practitioners and patrons, what was termed *ornato* might be lower down the aesthetic scale

than the 'formal purism' of beauty based on conformity with geometric laws, but in practice it was a widely desired objective.

In this chapter, as you may have noticed, I am conflating 'stone' and 'mineral': in the Renaissance both were commonly described as *pietre dure*, or hard stones, and many 'marbles' were in fact minerals. A workshop – the Galleria dei Lavori – to develop and teach the skills of creating decorative inlays using such materials was established in 1588 under the patronage of Ferdinando I de'Medici. Later known as the *Opificio delle Pietre Dure* it promoted a decorative style that flourished across Europe well into the nineteenth century and it is still active as a world-leader in the conservation of architecture and decorative art.

The focus of the work promoted by the Medici was on intricate decorations at the scale of furniture – tabletops and cabinets – more than large-scale architectural surfaces. It made extensive use of minerals and combined small-scale geometric arrangements along the lines of the Cosmati pavements with *opus sectile*-style work whose intricacy pushed the limits of both craftsmanship and materials. The Medici had representatives across Europe charged with acquiring fine materials, and the Opificio is reputed to have a stock of some six million pieces.

A superlative early example of the Galleria's work, now in the Metropolitan Museum in New York, was made for Maffeo Barberini, sometime between his becoming a cardinal in 1606 and being elected Pope in 1623. On the door of the central compart-

ment Orpheus is seen playing a *lira da braccio*, an early type of viol, charming animals and grotesque beasts with his music: the style is almost painterly. Other, more decorative panels display examples of animal cunning or folly based on Aesop's Fables; there is also a phoenix (which, according to classical myth lived for centuries then burnt itself to ashes, only to re-emerge from the fire), and a salamander whose frigid body, according to Pliny, rendered it impervious to fire. Identical vases with differing arrangements of flowers are placed at each side of the cabinet's front, while large panels with floral branches and colourful birds decorate the ends and top. The workmanship is of staggering quality and is thought to have been undertaken by German craftsmen who had settled in Florence.

The world of minerals and *pietre dure* we have reviewed in this chapter might seem a world apart from the technical and social developments to which modern architecture – my almost lifelong interest – was a response. But this is not entirely true. The Viennese architect Adolf Loos was a vigorous advocate of using large sheets of thin, highly figured stone – such as the green Cipollino marble that clads the lower two floors of the so-called Looshaus in Vienna (p.191) – to distinguish between surface cladding and underlying structure. Such natural materials, he argued, offered a new form of *modern* luxury, dependent not on expensive craftsmanship but on using industrialised quarrying and finishing to display their innate qualities.

Interest in the beauty of minerals, even the supposedly occult power of crystals, did not disappear entirely under the weight of scientific culture. Peter Behrens, for example, one of the key figures in the transition from Jugendstil to 'early modern' was much taken with the symbolism of crystals. At the opening of the Darmstadt Artists' Colony – a centre for the late-flourishing of Jugendstil (below) – in 1901, he presented a *Zeichen* or sign. It took the form of a crystal, and was offered as an emblem of transformation: just as carbon, under conditions of intense heat and pressure can metamorphose into diamond, so everyday life, he suggested, could be raised to a higher plane by art.

For the German section of the Exposition of Decorative Arts held in Turin in 1902, Behrens designed a cave-like vestibule with, as Stanford Anderson describes it, 'stucco walls and ceiling moulded into continuous yet hard, stalactitic surfaces.' In the ceiling was a stained-glass sunburst, and below a large volume of Nietzsche's *Also sprach Zarathustra* was displayed in a binding designed by Behrens, with a crystal as its decorative motif. For Behrens, crystals symbolised his vision of *life as art* and after the War they were adopted by a group of young German architects whose correspondence became known as the 'Crystal Chain Letters'.

The fascination with the clarity and purity of crystals informed thinking about the quintessential modern material, glass. When the most gifted architect of the new generation, Mies van der Rohe, was asked to contribute to the prospectus of the Association of German Mirrorglass Factories, he identified himself with this tradition by concluding that the 'purity' of modern materials reflected 'the luminosity of original beauty' – an echo, he said, of 'the profound words of St Augustine' that he quoted throughout his life: 'Beauty is the radiance of truth.'

In the German pavilion at the Barcelona international exhibition of 1929 Mies employed three types of glass – clear, milky-white (used to create a mysteriously glowing top-lit 'box') and green-tinted – and four different types of 'marble': Roman travertine (which has an almost magical ability to absorb and seemingly re-transmit sunlight), green Alpine marble, an ancient green marble from Greece and an 'onyx' mined from the Atlas Mountains in north Africa. The term 'onyx' is generic, commonly used to refer to the mineral chalcedony; in the Pavilion the material is the closely related carbonate mineral aragonite.

The Barcelona Pavilion was dismantled after the exhibition, but the original sources of the stones were scrupulously tracked down when it was reconstructed, and eventually re-opened, in 1986. Comparing the experience of walking around the reconstruction with the famous set of black and white photographs on which the Pavilion's reputation as one of the defining achievements of Modern architecture rested, I cannot help feeling that something has been lost. The differences appear slight, but seem to render the interior less atmospheric.

The wall of green-tinted glass framing the pool presided over by the statue of Dawn appears much more transparent than in the photographs, significantly reducing the interplay of reflections that Mies explored through carefully constructed perspectives. The book-matched veneer of the focal 'onyx' wall has a more assertive geometric pattern, culminating at one end in a complete diamond: in the original, the geometry resulting from the book-matching was less conspicuous and the pattern decidedly more chaotic. Mies was also careful to exclude the surrounding world, positioning shrubs in large pots where required, reinforcing the feeling of being immersed in an enchanted realm.

Following the success of his pavilion Mies was commissioned to build a house in Brno for the wealthy industrialists Fritz and Grete Tugendhat. He deployed a similar rich palette of materials, including exotic wooden veneers; velvet and shantung silk curtains; black, red and emerald leathers on chairs of his own design; and, as a principal element of the living space, another plane of aragonite, much thinner than in the Barcelona Pavilion. There is even, at the east end, a linear winter garden: too narrow to sit in, it creates, early on sunny mornings, a spectacular display of shadows on the white silk curtain that runs along its length. Listed in this way the house's materials bring to mind the ecstatic world of the *Hypnerotomachia*, and nature, fittingly, was to add her own unexpected magic. Like many silicate minerals, aragonite can present surprising changes of colour: lit by low sun in the late afternoon, the much thinner panel used in the house transmits light through translucent veins, transforming it into a glowing red. 'It was,' Mrs Tugendhat later noted, 'a joyous surprise even for Mies.'

Where, you might reasonably be wondering, is this excursion into an exquisite but rather particular corner of Modernist aesthetics leading? If it were merely a matter of luxurious materials artfully deployed it would hardly be worth the detour – Art Deco affords any number of examples of 'designer luxury'. Mies clearly relished the opportunity to use some of the finest materials the world could offer, but his intentions were far from those of a decorator and also ran counter to Modernist orthodoxy. In his seminal 1923 book *Vers une architecture*, Le Corbusier was scathing about 'the old-world timber beam', in which 'there may be lurking some treacherous knot,' and argued that 'natural materials, which are infinitely variable in composition, must be replaced by fixed ones', by which he meant 'homogeneous and artificial ones', such as steel, reinforced concrete and glass.

What Mies sought was not the stability of the artificial, but the 'infinite variability' of nature that the young Le Corbusier condemned. During the 1920s he became involved in a movement for the renewal of the Catholic church in which the Berlin-based priest and academic, Romano Guardini – considered by many the leading Catholic intellectual of the century – played a key role. In his book *Letters from Lake Como* Guardini pondered the challenge of reconciling Christian tradition and nature with modern life and technology. At one point he writes that 'there is a totally untouched nature, and the longing for it is itself a cultural phenomenon ... Nature is truly affecting only when it begins to be dwelled in, when culture begins in it. Piece by piece nature is formed. Man creates it in his own world, not only of a natural need, but with a deliberate purpose, serving spiritual ideas.'

What Mies was attempting to create in the Barcelona Pavilion and Tugendhat House was radically new, going far beyond the familiar modern trope of uniting architecture and nature through the 'flow' of 'interpenetrating spaces'. The buildings became a complex interplay of reflections, shadows, filtered views and surfaces animated by the 'inner life' of natural materials. Even the slender steel columns are caught up in the game. Cruciform in section and clad in polished chrome they are transformed into streaks of light, apparently incapable of doing serious structural work. By questioning traditional distinctions between nature and culture, building and site, structure and form, inside and out, surface and depth, Mies offered an architectural synthesis that was both conceptually complex and a visual delight. If, as I hope, my 'data from nature' are to find use in architecture, nothing built since these masterpieces seems to me to offer a more inspiring springboard for exploration.

7

CARLO
ALCITE

Discovered by two back-packing students in a lead-lined casket in the dank basement of the ruins of a remote Franciscan monastery in Liguria in 2012, Carlo Alcite's work is the most exciting art-historical discovery of the twenty-first century. Alcite seems to have been born around 1620 in Genoa, then a thriving centre of baroque painting, but did not work primarily in oils or any of the major genres of his time, focussing largely on etchings and drawings of landscapes. Many of his works suggest geological insights into the formation of land forms, while his occasional efforts with watercolour anticipate the topographic precision of the British eighteenth-century school.

Given the dearth of information about his life, and the poor condition of much of his paper-based *œuvre*, Alcite's work presents formidable interpretative and conservation challenges. One of very few documentary references to him that have so far come to light appears in the journal of his fellow Genoese artist, Orazio de Ferrari (1606–1657). It refers, tantalisingly, to a visit to '*un paese del nord di ghiaccio*' ('a northern country of ice'). In a forthcoming doctoral thesis entitled *Topografia e località: sito specificità gli acquarelli di Carlo Alcite* (Topography and locality: site specificity in the watercolours of Carlo Alcite) Allegra Monterubbianesi will argue persuasively that Alcite was the first major artist to visit Iceland. Following exhaustive field-work she has been able, for example, to link several studies to the landscape of the south-east coast, west of Fagurhólsmyri. While the timing of Alcite's epic journey is uncertain, the experience of the Icelandic landscape, where geological processes are laid bare more vividly than almost anywhere else on Earth, clearly played a vital part in the development of his mature style.

Monterubbianesi has also been able to offer convincing evidence that Alcite travelled through Holland and Britain on his way north. This is

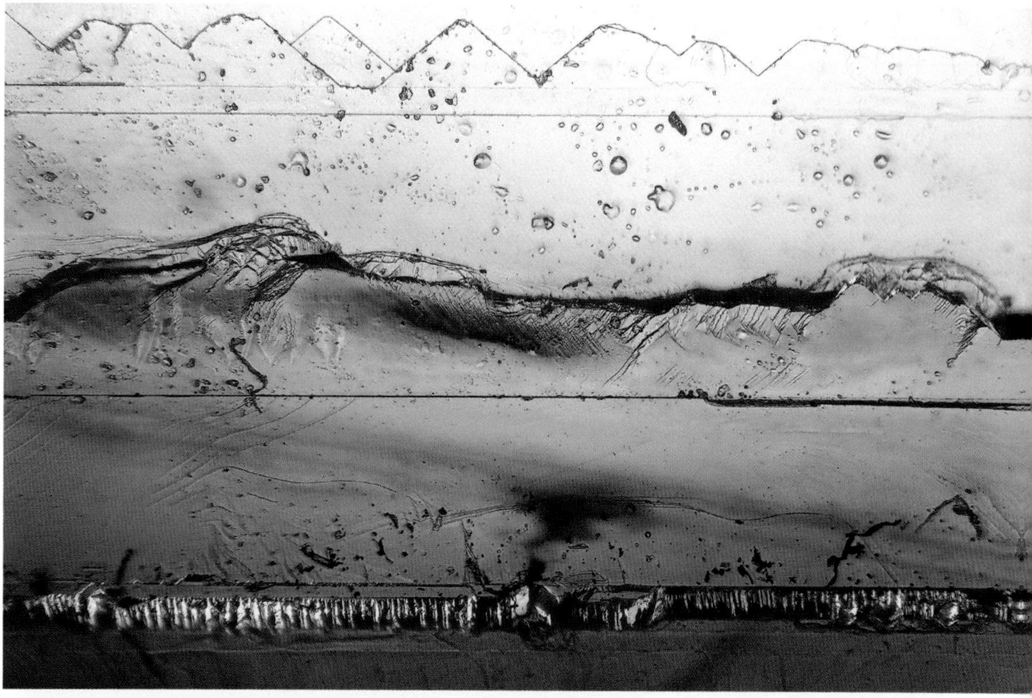

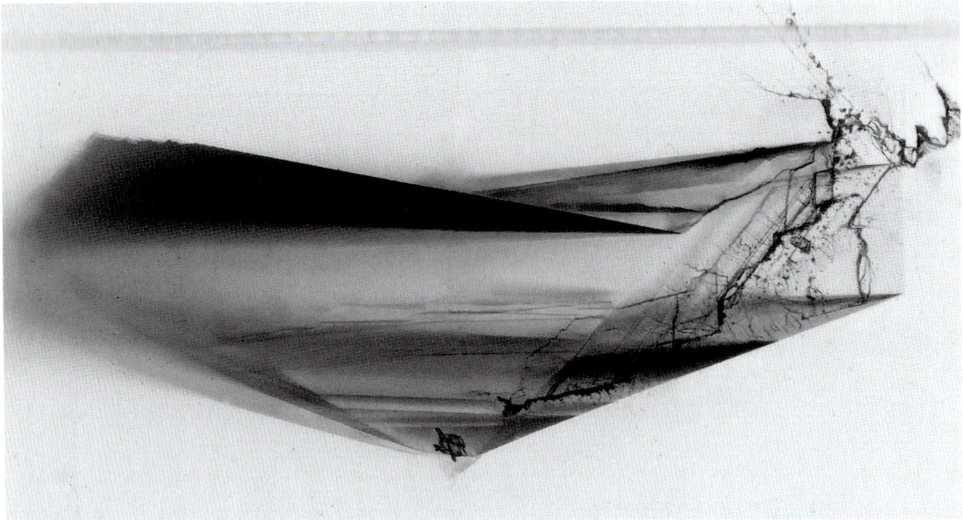

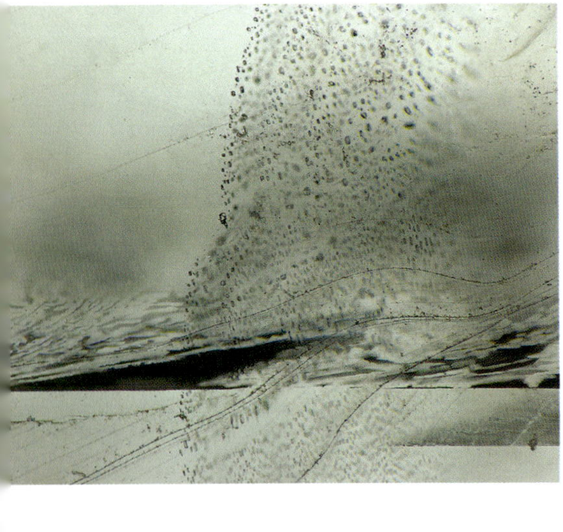

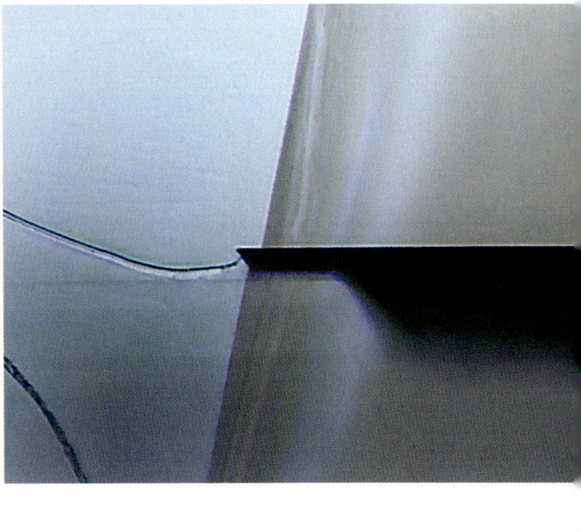

especially fascinating because the only artist whose work bears a close stylistic correspondence to Alcite's is Hercules Segers, arguably the most original of the first generation of Dutch landscapists. The subject of a major exhibition at the Rijksmuseum in 2016, which then travelled to the New York Metropolitan Museum, Segers's work was previously known largely by specialists. He was, however, greatly admired in his day, notably by Rembrandt who built up a substantial collection of his work.

In Italy Alcite seems to have moved more widely in scientific than artistic circles. One not especially appealing drawing shows several vertical 'smudges': too large to be conifers, what might they represent? The answer, I believe, is to be found in similar studies of smoke emanating from below the ground. We now know that these 'everlasting fires', as they were called at the time, were burning underground in seams of coal or petrifying wood. Four sites where 'fumes are seen to issue from the earth' are identified on a map of the area around Rosano and Dunarobbo in Tuscany, drawn by Francesco Stelluti in 1637 (Royal Library, Windsor, RL 25688). The map was commissioned by Cassiano dal Pozzo for a monumental documentation of nature now known as his 'Paper Museum'. A close friend of the leading French painter Nicolas Poussin, dal Pozzo was also a noted correspondent of Galileo, the most distinguished member of the Lincean Academy founded by Federico Cesi in 1603. The Linceans, one of the most important early-modern associations of scientists that acted as a model for Britain's later Royal Society, believed that the study of nature should be based on direct observation, not filtered through Aristotelian philosophy. Galileo made and presented to them the first microscope in Italy, and following Cesi's early death in 1630, dal Pozzo acquired an extensive collection of micrographs made by the Linceans and incorporated them into his Paper Museum.

I have dwelt on these connections for two reasons: firstly, because the empirical scientific spirit pioneered by the Lincean Academy is, I believe, vital to an understanding of Alcite's work; and secondly, because the complex structures and relentless depiction of detail in many of his drawings may well owe something

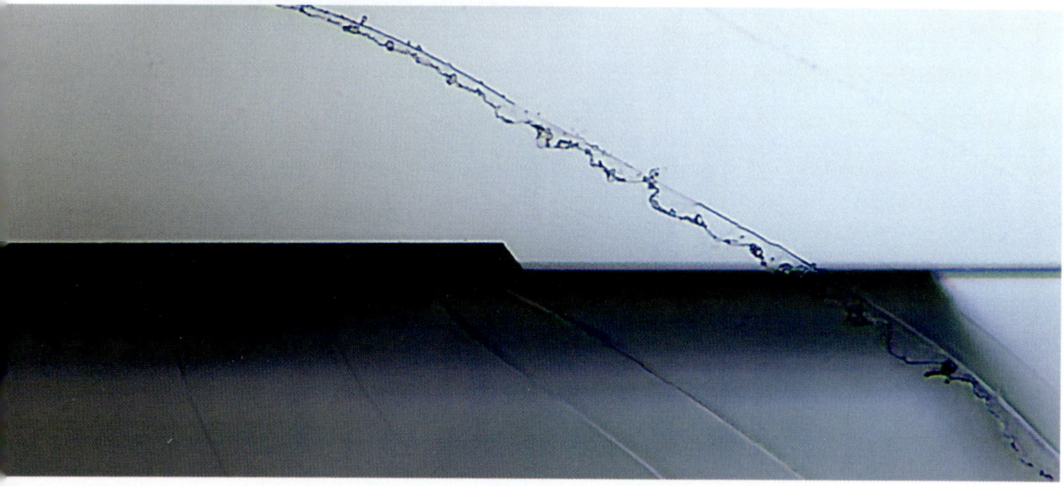

to micrographic studies of crystals, transposed to the larger scale of landscapes. If, as seems likely, Rembrandt introduced Alcite to the work of Segers, he might well have introduced him also to the leading Dutch scientist Christiaan Huygens, son of his patron Sir Constantijn. The younger Huygens was fascinated by the optical properties of what was then called Iceland crystal, a clear form of calcite we know today as Iceland spar. On his third visit to England in 1689 Huygens is known to have discussed its optical properties – it is a natural polariser – with Isaac Newton, and a crystal such as this would surely have fascinated Alcite.

What is abundantly clear from his work is that Alcite was drawn to manifestations of natural *processes* as much as *appearances* – as, before him, were Leonardo da Vinci and the French ceramicist of genius, Bernard Palissy. For Alcite, the complex systems of order to be found in crystals – of micro-strata and systematic cracks, for example – would have been a revelation, offering microcosmic intimations of similar, large-scale systems beneath the Earth's surface. Although I believe Alcite should be firmly associated with the emergence of empirical science, it is worth remembering that Rosicrucianism, with its claims to esoteric knowledge about the Earth based on Kabbalah, Hermeticism, alchemy, and mystical strands of Christianity, was thriving in the early seventeenth-century. The idea of the microcosm was promulgated by Robert Fludd, a defender if not follower of Rosicrucianism, and expressed in two diagrams, one resembling the familiar Vitruvian Man, the other a more intriguing evocation of links between mind and universe (p.179).

Many of Alcite's meticulous studies of rock formations evoke a feeling for what we now call geomorphology, and even those that appear non-figurative can be read as imaginative

renderings of ephemeral processes – freezing, precipitation, wind, crystallisation. Falling rocks or tiny crystal fragments – depending on how your read the scale of certain images – are frequently scattered across the pictorial field: frozen in free-fall, they evoke a kind of 'universal weather'.

The sense of flux that pervades Alcite's work was doubtless inspired by the Classical idea of *natura naturans* that lingered on through the Renaissance and was revived by Alcite's contemporary, the philosopher Baruch Spinoza. Regardless of any specific influence from Spinoza, the idea of nature as self-generating, dynamic, and animate – as process rather than form – was widely espoused, informing, for example, Leonardo's celebrated studies of turbulence provoked by childhood memories of a cataclysm that devastated Tuscany in 1456. This was vividly described by Machiavelli in his *Istorie Fiorentine*, and interest in such disasters was widespread, as confirmed by the recently rediscovered *Book of Miracles* compiled in the Swabian Imperial Free City of Augsburg around 1550. In Alcite, all such cataclysms – and they are a recurring theme – appear natural, part of an ongoing cycle of creation and destruction, not the arbitrary interventions of an angry God.

Alcite's vision of 'nature naturing' extended to the conception of architecture as an outgrowth of a given site condition, not an arbitrarily imposed construction. He made numerous architectural

studies and they range from vast platforms reminiscent of Mayan temple complexes to intricately interwoven topography and planes that can appear strikingly modern – the example shown here might almost be a study for the Portuguese architect Alvaro Siza's celebrated tidal pools on the coast at Leça de Palmeira in Portugal (1959-73). A recurring architectural theme are viaduct-like structures that cut through the terrain like formalisations of geological faults, but might well have been motivated by the restoration of a Roman viaduct there, the major public work in Genoa during his lifetime: for the modern viewer these studies appear uncannily like proposals for works of Land Art. Alcite's architectural projects were generally represented from above, perhaps reflecting a fascination with the celebrated aerial view of Genoa in Domenico Fiasella's celebrated *Madonna della Cittá*, painted c. 1640 and then the city's best-known work of art; it is now in the Chiesa della Pietà dei Turchini in Naples.

In nearly all Alcite's architectural *capricci* we see rock outcrops being transformed into architecture. Such processes of transformation run through the Renaissance and Baroque periods. John Ruskin, for example, suggested that the rustication of the 15th-century Pitti Palace alludes to natural strata revealed in a quarry, while a late Baroque masterpiece like the Trevi Fountain offers a vivid image of a similar dialogue between nature and architecture, formation and erosion.

Alcite's drawings are also replete with what the art historian Dario Gamboni calls 'potential images' – suggestions of faces, animals or other natural forms half-concealed as part of the topography. Gamboni has traced their lineage back to Classical times and a celebrated example that Alcite could well have known is the scowling face in a cliff in Albrecht Dürer's *View of Arco* (1495). With Alcite, however, the case is more complex: so abundant are such 'potential images' that it seems inconceivable that they

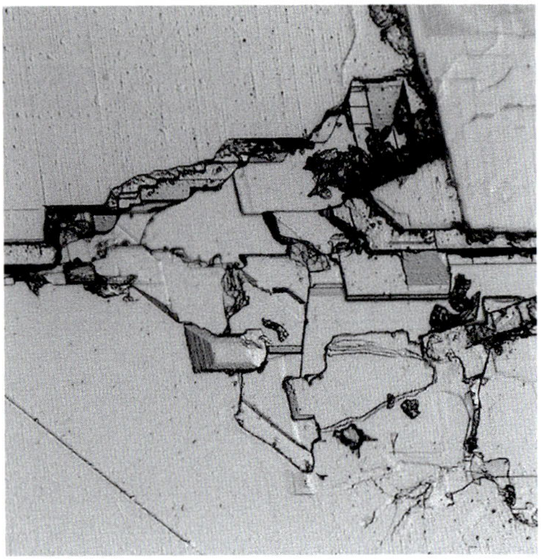

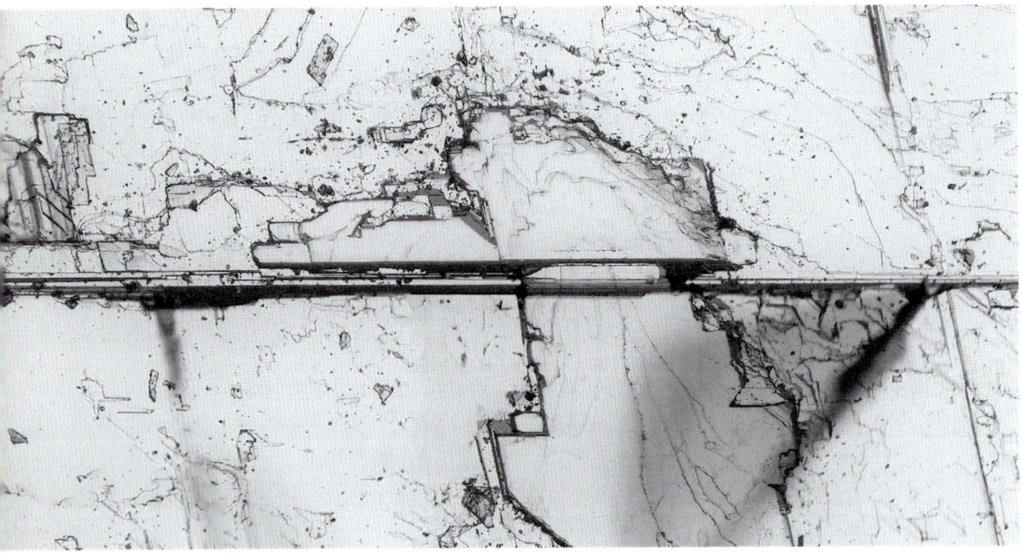

were all intended: they may be better thought of as pareidolia, evocative accidental configurations like Rorschach blots.

Although the vast majority of Alcite's landscapes show land forms stripped of vegetation or other forms of life, he seems to have been fascinated by dead or dying organisms washed up on beaches (p.215, top). Some suggest the process of fossilisation, a lively topic of discussion among artists since the early sixteenth century (Leonardo famously drew the then fashionable women's hairstyle modelled on the spiral of an ammonite), and they also recall Bernard Palissy casting 'animal impressions' from snakes, frogs, lizards and other creatures: the process is discussed at length in his *Discours admirables* of 1580. As Hannah Rose Shell has observed, Palissy's rustic ceramics 'compressed both nature and time into clay': working a century later, Alcite inscribed into paper the moment of death of tiny, neglected life-forms as magisterially as he showed that the seemingly adamantine large-scale structures of the earth are subject to inexorable processes of transformation and decay.

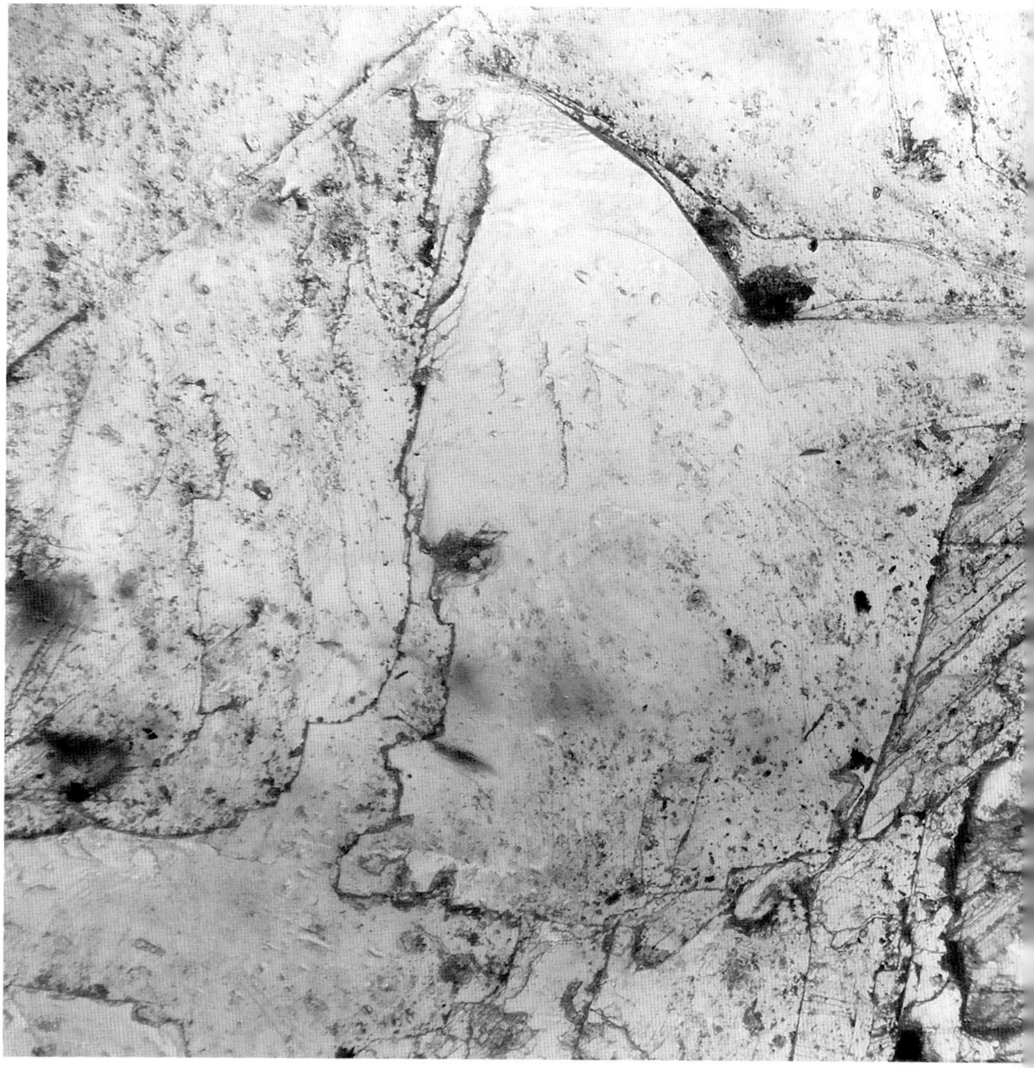

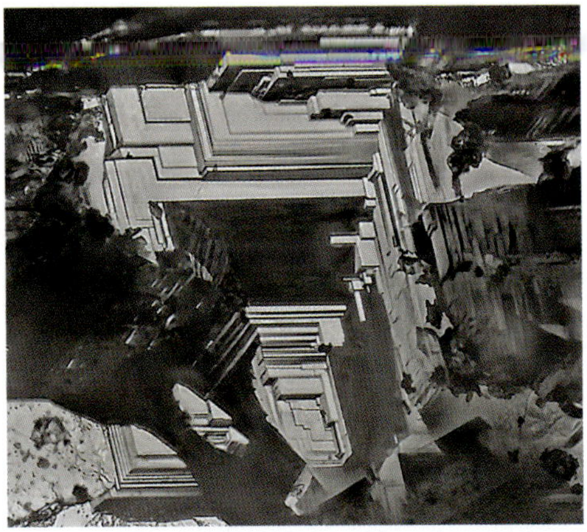

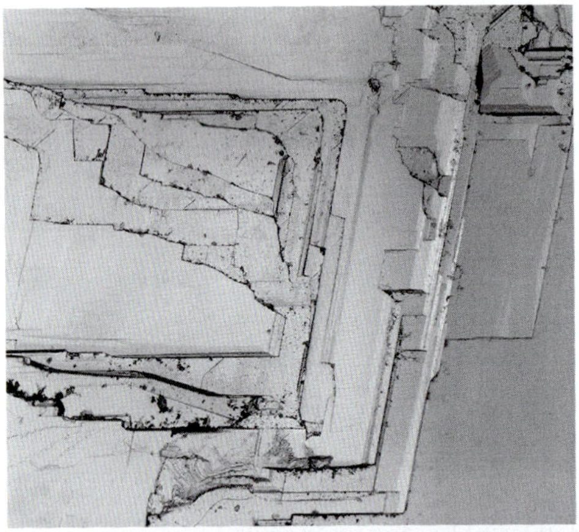

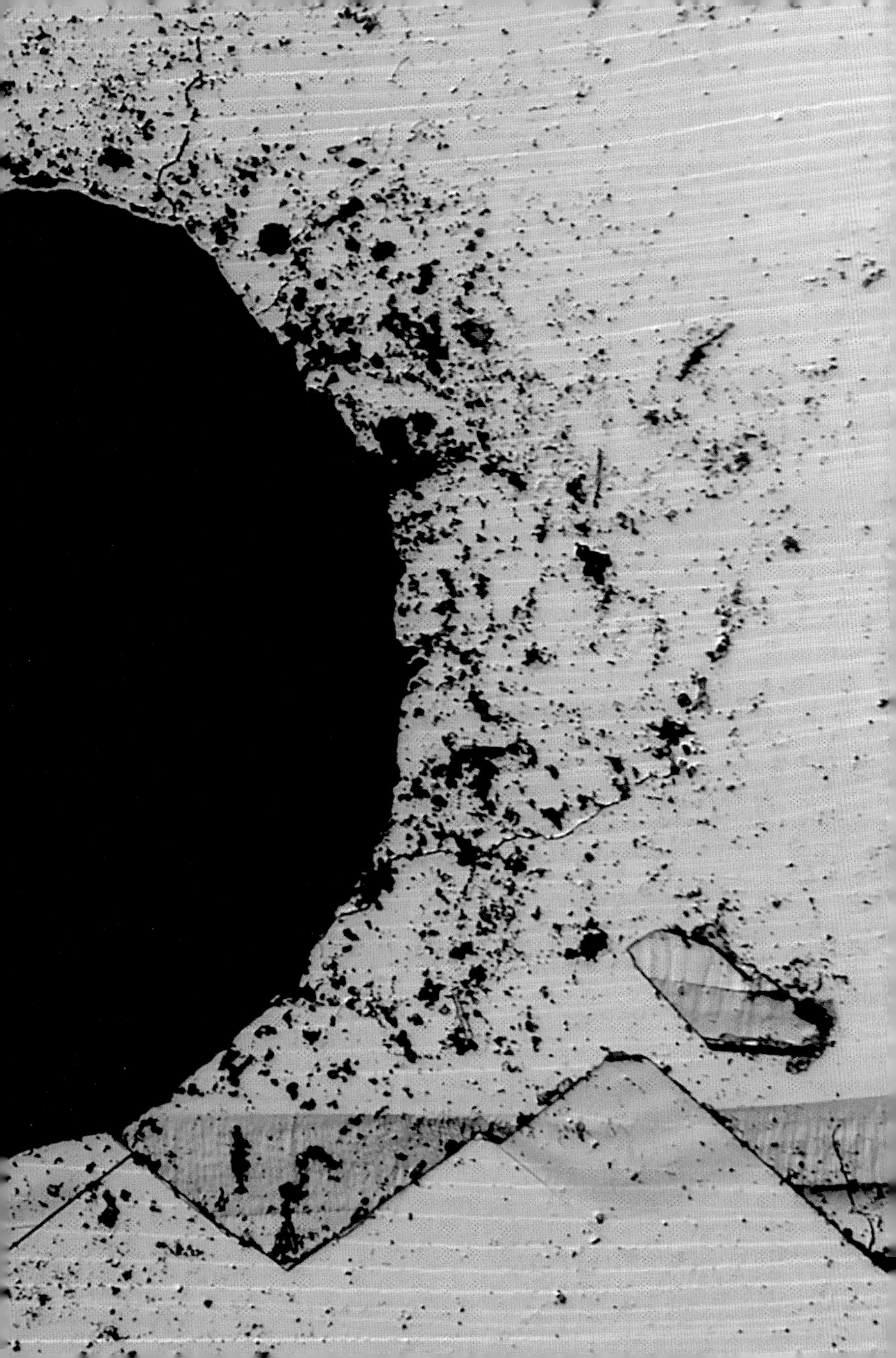

CON-
CLUSION

When I began these adventures with 'data from nature' the word 'app' had not entered general usage, mobile phones were devices with which to make phone calls and send text messages, and the launch of the iPad was still nearly seven years into the future. Fifteen years on, the means to embrace the Digital Revolution as creative producers, not largely passive consumers, are widely affordable. But while the apps I am developing are clearly unthinkable without digital media, the values they embody go back to my student days. In 1971, the year I went up to university, Ivan Illich published *Deschooling Society*, advocating 'educational webs' in place of formal institutions, and arguing that 'most learning is not the result of instruction. It is rather the result of unhampered participation in a meaningful setting.' Molly's World and, in a more focused way, CreateForAll, will, with luck, evolve into just such settings – manifold, multi-faceted and continually expanding.

Two years later, in *Tools for Conviviality*, Illich advanced the idea of 'enabling tools which foster self-realization' and suggested that people 'need not only to obtain things, they need above all the freedom to make things among which they can live, to give shape to them according to their own tastes.' In the same spirit, Lee Felsenstein, a pioneer of the Personal Computer, suggested that 'if work is to become play, then tools must become toys': in many ways they have, but making the most of their creative potential requires a shift in values, away from defining ourselves by what we consume.

Thanks to social media, 'personalised' advertising and a growing army of online 'influencers', the pressures to conform to consumerist norms are becoming ever more intense. In 2018, $630 billion was spent globally on advertising, and as long ago as 1929, in a conference convened by President Herbert Hoover, parents were advised to take their children shopping to teach them that their personalities 'can be expressed through things.' As the environmental and social consequences of a global economy based on ever-growing consumption become inescapable, change must come. But governments seem incapable of imagining radical alternatives, apparently believing – against all the evidence – that with more efficient use of energy, recycling, a greater emphasis on renewable resources, and tweaks to the taxation system to redress the most extreme divisions of wealth in human history, something close to 'business as usual' will still be sustainable.

What we are experiencing as the Digital Revolution takes hold echoes some of the concerns of the great Victorian critic and thinker, John Ruskin, whose critique of the Industrial Revolution grew from his belief that it was turning workers into machines, denying them the opportunity to fulfil their potential through creative work.

The disparate chapters of this book share two themes: the versatility of 'data from nature' and the endless variety of natural and human creativity. I am thinking not only of the innate creativity of children, nor just of the potential that can be released by CreateForAll, but of the creativity of matter that pervades inorganic as much as organic nature. From the vast and immensely slow rock cycle we explored in chapter 1, to the exquisite manifestations of processes of crystallisation captured with a microscope, nature is inexhaustibly creative. We encountered an ancient recognition of this in the phrase *natura naturans*, of nature as the generator of new forms, that informed the thinking of ancient Romans such as Pliny and Lucretius, and was revived in the Middle Ages. The Greeks had a similar idea, encapsulated in the word *phusis*, which denoted processes of formation and deformation, of a perpetual motion through which things collide and forms come into being. Looking at the exquisite pattern (opposite) created by manganese molecules circulating randomly in a film of water between the bedding planes of limestone, I can just about imagine how it might come about through the process scientists describe as Diffusion Limited Aggregation – but the result still seems little short of miraculous.

Such patterns make one ponder taken-for-granted distinctions between living and non-living matter, and just such a questioning is proposed by advocates of what has become known as the 'material turn' in critical theory. At the end of her recent book *Vibrant Matter*, the political theorist Jane Bennett sets out a Nicene Creed for this new world-view, professing her belief in 'one matter-energy, the maker of things seen and unseen.'

Writing in the 1930s the critic and philosopher Walter Benjamin expressed the hope that 'optical instruments' like the camera and microscope would enable adults to recapture the sense of wonder in the face of nature that is typical of childhood. I don't think I ever entirely lost that sense, but delving into the inner life of minerals and other materials has given it an unexpected and overwhelming vividness. 'Vibrant matter' could hardly have more potent ambassadors than minerals and I cannot express this more evocatively than by recalling my former neighbour's instinctive reaction to the 'cosmic' image of orbicular jasper (pp.14-15), 'it's made like us,' or a young girl's response to mineral-colouring her drawing of a mushroom, 'it feels alive now.'

It seems to me only a small jump from these instinctive reactions to the aesthetic ideals of Ruskin, in particular to what he called 'living design'. Vital to this were 'changefulness' – for which the variety of nature was his model – and 'savageness', a quality that flowed from allowing less gifted craftsmen to work free of imposed designs. In Ruskin's sense (and perhaps in others!) children's work is eminently 'savage'; but it can yield things as evocative as the drawing of Molly that has become our emblem.

Ruskin based his ideal of 'living design' on a romanticised view of Gothic and used it as a whip with which to scourge the 'mechanical' world he saw flowing from the factories spawned by the Industrial Revolution. Dominated by commercialism, the Digital Revolution as we are experiencing it in its early stages seems to me to be similarly flawed, enslaving so many not to machines but to a range of all too familiar temptations, from violent computer games to social media and the latest colour of makeup. But for all the current worries about the impact of digital media, especially on children, the promise of radically new possibilities remains. I hope that the adventures presented in this book will in time offer a widely accessible stepping stone along the path towards genuinely personal ways of learning and making.

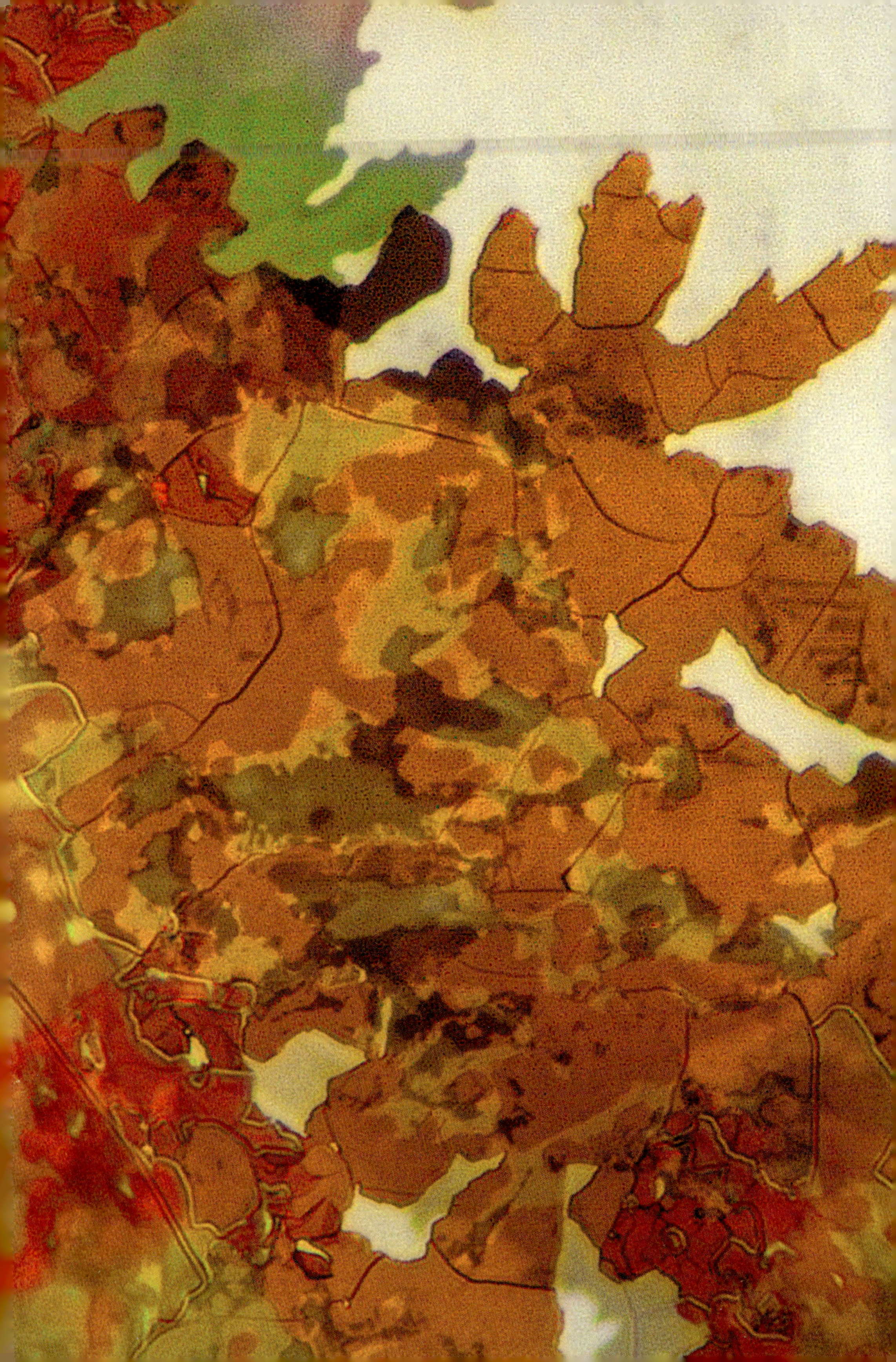

DATA FROM NATURE
NATURE, DIGITISATION AND DESIGN

Published by
LIStLab
info@listlab.eu
www.listlab.eu

Author
Richard Weston

Art Direction & Production
Blacklist Creative, BCN
blacklist-creative.com

ISBN 9788832080025

Printed and bound in the European Union
2019

series BABEL INTERNATIONAT

Promotion and distribution:
- Italy
Messaggerie Libri, Spa, Milano
assistenza.ordini@meli.it
amministrazione.vendite@meli.it
- Europe and International
ACC Book Distribution Ltd, UK
uksales@accartbooks.com
- China, Japan & South-East Asia
SendPoints, China
sales@sendpoints.cn

LIStLab is an editorial workshop, based in Europe, that works on contemporary issues. LISt Lab not only publishes, but also researches, proposes, promotes, produces, creates networks.

LIStLab is a green company committed to respect the environment. Paper, ink, glues and all processings come from short supply chains and aim at limiting pollution. The print run of books and magazines is based on consumption patterns, thus preventing waste of paper and surpluses. LISt Lab aims at the responsibility of the authors and markets, towards the knowledge of a new publishing culture based on resource management.